Computational Molecular Science and Engineering Forum 2013

Core Programming Area at the 2013 AIChE Annual Meeting: Global Challenges for Engineering a Sustainable Future

San Francisco, California, USA
3-8 November 2013

ISBN: 978-1-63439-035-4

Printed from e-media with permission by:

Curran Associates, Inc.
57 Morehouse Lane
Red Hook, NY 12571

CURRAN ASSOCIATES INC. proceedings.com

Some format issues inherent in the e-media version may also appear in this print version.

Printed by Curran Associates, Inc. (2014)

AIChE
3 Park Avenue
New York, NY 10016-5991

Phone: (203) 702-7660
Fax: (203) 775-5177

www.aiche.org

Additional copies of this publication are available from:

Curran Associates, Inc.
57 Morehouse Lane
Red Hook, NY 12571 USA
Phone: 845-758-0400
Fax: 845-758-2634
Email: curran@proceedings.com
Web: www.proceedings.com

TABLE OF CONTENTS

(49a) Identifying Optimal Active Sites for Heterogeneous Catalysis By Metal Alloys Based On Molecular Descriptors and Electronic Structure Engineering

Xin, H., University of Michigan
Holewinski, A., University of Michigan
Nikolla, E., Wayne State University
Linic, S., University of Michigan
Originally presented on: 11/4/2013 8:30:00 - 8:50:00

While one metal might be chemically inert for a particular catalytic transformation, its first neighbor in the periodic table might be overly reactive and equally inefficient. The main reason of this is that neighboring metals in the periodic table exhibit very large differences in binding energies of various adsorbates (as large as ~1 eV (~100 kJ/mol)). Simply put, the periodic table does not provide enough flexibility in identifying optimal catalysts. One avenue to address this problem is through fine-tuning the chemical reactivity of surface atoms in metals by changing their local coordination, for example by creating metal alloys. The phase space of alloys is immense, and it is almost impossible to screen through even a small fraction of potential alloys using experimental studies or even quantum-chemical calculations on model systems.

In this talk we will discuss our work on developing predictive models of chemical reactivity for alloys based on detailed understanding of the molecular transformations that govern catalytic processes. More specifically, we analyzed how the change in the local coordination of metal surface atoms due to alloying changes the local chemical reactivity of the metal site. The central finding of our studies is that it is possible to reliably predict the change in the local electronic structure of an active site induced by the formation of an alloy, and the change in the local chemical reactivity, based on very simple analytical models which employ only easily accessible physical characteristics of the elements that form the alloy (mainly their electronegativity, distance between neighboring atoms, and the spatial extent of metal orbitals). We show how these predictive structure-reactivity relationships can be employed to rapidly screen through large libraries of alloy compositions and structures to identify optimal alloy catalysts. We describe our findings by focusing on an example of the design of optimal Pt alloy electrocatalysts for the oxygen reduction reaction (ORR) in fuel cell cathodes. The model permits rapid screening through an enormous phase space of alloy structures and compositions using analytical expressions instead of expensive quantum-chemical calculations. Since the model is grounded on validated theories of chemisorption on metal surfaces, it can be used for the identification and targeted manipulation of the critical electronic structure descriptors of catalytic activity and propose how these features can be obtained based on electronic structure engineering.

1. **H. Xin**, N. Schweitzer, E. Nikolla, S. Linic, J. Chem. Phys., 132, 111101, (2010)
2. N. Schweitzer, **H. Xin**, E. Nikolla, J.T. Miller, S. Linic, Top Catal, 53, 348, (2010)
3. **H. Xin**, S. Linic, J. Chem. Phys., 132, 221101, (2010)
4. **H. Xin**, A. Holewinski, S. Linic, ACS Catal., 2, 12, (2012)
5. **H. Xin**, A. Holewinski, N. Schweitzer, E. Nikolla, S. Linic, Top Catal, 55, 376, (2012)
6. A. Holewinski, **H. Xin**, E. Nikolla and S. Linic, Curr Opin Chem Eng., accepted (2013)

(49b) A Computational Evaluation of Steady-State Bifunctional Catalysis

Doan, H. A.,
Grabow, L. C., University of Houston
Originally presented on: 11/4/2013 8:50:00 - 9:10:00

Significant improvements in computer hardware, increasingly efficient and accurate density functional theory (DFT) codes, and the use of scaling and Brønsted-Evans-Polanyi relations in heterogeneous catalysis, have enabled us to predict optimal materials for various catalytic processes computationally. However, as the best catalysts are not always practical due to their high cost, it is essential to continue the discovery new and affordable materials with high catalytic activity. In the literature, several examples of high performance catalysts with multiple site-specific functionalities at steady-state have been reported [1-4]. The most common systems consist of two distinct sites that catalyze different reaction steps independently and are often referred to as "bifunctional catalysts". Using DFT and microkinetic modeling, we aim to understand the mechanisms that are responsible for bifunctional activity and provide a computational framework for screening of bifunctional catalysts. Our results indicate that there are theoretical limits for the achievable activity improvement and bifunctional catalysts do not necessarily outperform single-site catalysts. More specifically, for CO oxidation on bimetallic systems we found that the overall activity is not significantly altered when bifunctional catalysts are considered, but equally active bifunctional catalysts may be tailored from less active and cheaper components.

1. Kotobuki M, Watanabe A, Uchida H, et al. (2005) Journal of Catalysis 236:262–269.
2. Thinon O, Diehl F, Avenier P, Schuurman Y (2008) Catalysis Today 137:29–35.
3. Neurock M (2010) Industrial & Engineering Chemistry Research 49:10183–10199.
4. Kyriakou G, Boucher MB, Jewell a. D, et al. (2012) Science 335:1209–1212.

(49c) Decoupled Scaling Relationships of Atomic and Molecular Adsorption Energies On Transition-Metal Carbide Catalysts

Michalsky, R., Brown University
Zhang, Y.,
Medford, A., Stanford University
Peterson, A. A., Stanford University
Originally presented on: 11/4/2013 9:10:00 - 9:30:00

Currently in the United States, natural gas is the primary feedstock for the production of many major chemical commodities such hydrogen and ethylene. The sustainable synthesis of these chemical energy carriers via the catalytic reduction of abundant resources such as carbon dioxide and water with renewable energy and without fossil fuel requires the development of efficient and inexpensive catalysts for large-scale production. The design of transition metal or metal alloy catalysts is often limited by the scaling of adsorption energies of reactants and reaction products. This work uses electronic-structure calculations to present how transition-metal carbide surfaces can decouple the scaling of hydrogen, carbon and oxygen adsorption energies - thereby increasing the degrees of freedom for the rational design of novel catalytic materials. Throughout the 12 studied transition metal carbide surfaces, carbides show an oxophilic departure from the scaling relations of adsorption energies on transition metal surfaces and a carbophobic shift of carbon adsorption energies by 0.30 to 1.8 eV relative to the (211) surface of the parent metal. Given a highly hybridized and delocalized electronic structure, the metal-projected d-band centers correlate with the reported adsorption energies. However, the preference of carbides for oxygen adsorption, site preferences of adsorbates, and the tendency of a surface to form C-C bonds between the lattice carbon and adsorbed carbon may be understood due to shifts in the sp-band center and an increasing degree of electron localization with decreasing electronegativity of the metal. Correlation of atomic adsorption energies with relevant molecular adsorbates such as CO, HCO, COOH or HCO2 as well as the activating or passivating influence of carbon vacancies or oxygen adatoms on adsorption energies respectively will be demonstrated.

(49d) Elucidating Adsorption Mechanisms On Oxides By Comparing Adsorption On FCC Metals and Reduced B1 Monoxide Surfaces

Xu, Z., Carnegie Mellon University
Kitchin, J. R., Carnegie Mellon University
Originally presented on: 11/4/2013 9:30:00 - 9:50:00

In the past decade, we have been able to use density functional theory to connect adsorption on metal surfaces with their electronic structures. This has allowed us to use calculated electronic structures to both explain adsorption phenomenon and create predictive models to accelerate materials design. However, our understanding of oxides pales in comparison. This is in large part due to both their diverse atomic and complex electronic structures. In an effort to understand the relationship between oxides' electronic structure and their adsorption properties, we compare side by side the bulk thermodynamics and surface adsorption properties of 3d transition metals in fcc structure with their corresponding B1 transition metal monoxides. Understanding that the B1 monoxide structure constitutes two interpenetrating face centered cubic lattices of cations (metals) and anions (oxygen), we construct a thermodynamic path from fcc metals to B1 monoxides through the expansion of the metal lattice and the insertion of an oxygen fcc lattice. We then studied how these structural transformations affected both the adsorption energies and electronic structure. We found that, in general, the expansion of the lattice corresponds to a weakening of the bond, which maximizes for metals with half filled d-orbitals. Interestingly, the introduction of oxygen into the expanded lattice strengthens the bond, which also maximizes for metals with half filled d-orbitals. The net effect is that the trends for the adsorption on the (111) surface of fcc metals mirrors that of their corresponding (111) B1 monoxide surfaces. We also concluded that irrespective of the material, the key descriptor of the oxygen adsorption energy is the center of the adsorbate oxygen p-band, which we then used to interpret the relationship between adsorption energies and bulk electronic structures. In the case of the metal, the oxygen p-band center is inherently tied to the d-band center, but for the oxide, the oxygen p-band center is tied to the properties of the d-band bonding, non-bonding, and bonding orbitals that are created when oxygen is inserted into the expanded metal lattice. We discuss possible structural and electronic structure descriptors of reactivities on oxide surfaces.

(49e) Reactivity of Acetaldehyde On CeO2(111) Surfaces and the Roles of Oxygen Vacancies

Xu, Y., Louisiana State University

Originally presented on: 11/4/2013 9:50:00 - 10:10:00

Ceria is a widely used catalytic and functional catalyst support material, well known for its ability to store oxygen and change oxidation state. There is a growing body of evidence that the surface reactivity of ceria toward organic molecules can vary significantly with the extent of reduction. In this study we use acetaldehyde as a probe molecule. First-principles density functional theory (DFT) calculations have been performed in close collaboration with surface characterization techniques to elucidate the roles of oxygen vacancies in redox reactions on the $CeO_2(111)$ surface. A microkinetic model based on proposed surface intermediates and their calculated energetics has been constructed and yields close agreement with experimental temperature-program desorption results. Acetaldehyde desorbs without reaction from the stoichiometric $CeO_2(111)$ surface near 200 K. When the surface is partially reduced, acetaldehyde loses its carbonyl bond character and forms C-O coupled dimers at very low temperatures. Increasing temperature to 360 K leads to the decomposition of this dimer species into acetaldehyde, which desorbs immediately, and another surface species, which has been conclusively identified to be the enolate form of acetaldehyde (CH_2CHO) and has not been captured on ceria surfaces previously. This enolate species can recombinatively desorb, again as acetaldehyde, at 550 K. Our findings demonstrate that surface oxygen vacancies are key to activating acetaldehyde and stabilizing it for further reactions, and that the dominant surface reaction pathway can change as a function of vacancy concentration. This work has relevance to the conversion of biomass-derived oxygenates because enolate species are key intermediates in C-C coupling reactions including aldol condensation.

(49f) Dopant Selection Guided By Sabatier's Principle: The Example of Doped Ceria for Promoting the Two-Step Water Dissociation

Botu, V., University of Connecticut
Mhadeshwar, A. B., University of Connecticut
Ramprasad, R., University of Connecticut
Originally presented on: 11/4/2013 10:10:00 - 10:30:00

Use of *ab-initio* methodologies, based on Density Functional Theory, in describing surface science phenomenon has allowed for an increasing predictive ability of materials. In the case of H_2 synthesis, cerium based oxides have garnered significant interest given their oxygen buffering capability that facilitates the creation of active surface defect sites (vacancies). Tailoring the activity and selectivity of these materials towards low temperatures is paramount to improving reactivity. We started investigating the fundamentals governing the surface phases of cerium (IV) oxide under a range of environments. In the case of a pure ceria system in an O_2 environment, it was found that T > 1500 K or $P_{O2} < 10^{-13}$ atm are required for any appreciable surface reduction. However, exposing the surface to a highly reducing environment such as CO or H_2 could help circumvent these harsh conditions. Given the aforementioned observations we sought to alter surface reducibility by doping with various alkali and transition metal elements. Doping ceria had an effect on the surface vacancy formation; low valence dopants favor defect formation, whilst high valence dopants suppress it. Armed with the insight dopants have on surface vacancy formation, their corresponding impact on the consecutive dissociation of water at these sites was studied. Using a carefully strategized multi-step screening approach abiding with the Sabatier's Principle, we were able to identify a subset of dopants that would promote surface reactivity towards H_2 synthesis. Dopants such as Sc, Au, Co, Pd, La, Y, Hg, Mn, Zr, Cr, and Fe possessed a critical balance in improving surface reducibility while sustaining the dissociation of water, hence are promising candidates in enhancing the two-step water dissociation process.

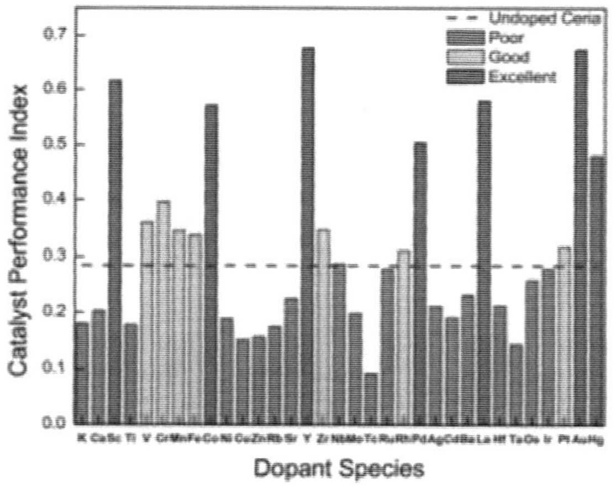

Figure 1: Measure of the Catalyst Performance Index for the class of dopants considered within ceria

(49g) Fundamental Insight Into the Unique Reactivity of NO Oxidation On the Mn-Ce Mixed Oxide

Wang, X., Tianjin University
Wang, J.,
Shen, M.,
Mei, D., Institute for Integrated Catalysis
Originally presented on: 11/4/2013 10:30:00 - 10:50:00

NO oxidation is the key step for NOx trapping, soot combustion under NOx/O_2, and NH_3-SCR. Recently, MnOx-CeO_2 catalyst has been found to be a promising catalyst to substitute the precious metals (e.g., Pt) for NO oxidation for its high activity. To understand the synergistic effect of MnO_2/CeO_2 catalyst, NO oxidation over CeO_2, β-MnO_2 and MnO_2/CeO_2 catalysts was systematically investigated using DFT+U method. For the perfect CeO_2 (111), we found that the NO adsorption energy is strong (-1.14 eV), which suggests that the extraction of lattice oxygen from CeO_2 (111) to form NO_2 is preferred. However, NO_2 desorption from the CeO_2 (111) is very difficult, which needs 2.51 eV. On the other hand, the adsorption of NO over the β-MnO_2 (110) surface is also strongly exothermic with an adsorption energy of -2.50 eV. Compared with CeO_2 (111), the NO_2 desorption on the β-MnO_2 (110) surface is much lower (0.48 eV). This is consistent with the experimental observations that MnOx sample exerts obviously higher NO oxidation activity than CeO_2. For MnO_2/CeO_2, the interaction between MnO_2 and CeO_2 (111) surface forms the interface Mn—O—Ce structure. The NO adsorption energy increases as the Ce site (-1.10 eV) ≈ the Mn site (-1.00 eV) < the interface site (-0.47 eV). While the NO_2 desorption energy increases as the interface site (0.48 eV) < the Mn site (1.02 eV). This implies that the NO_2 desorption on MnO_2/CeO_2 has been significantly facilitated by the interface sites, which only requires the lowest NO_2 desorption energy. To complete the catalytic cycle, the oxygen vacancies generated by the NO_2 desorption have to be replenished. Our calculations show that the adsorption energy of single oxygen atom over MnO_{2-x} is -0.33 eV while it is -1.88 eV at the interface site over MnO_{2-x}/CeO_2 catalyst. This indicates that the MnO_{2-x}/CeO_2 catalyst regenerates the active lattice oxygen much easier than the MnO_{2-x}. Our theoretical modeling results elucidate the experimental observation well.

(56a) Computational Approach of Thermodynamic Fragmentation Applied to Formulating Badly Soluble Actives

Fraaije, J. G. E. M., Culgi BV
Bulacu, M., Culgi
Gracia, R. S., Culgi
Nath, S., Culgi
Originally presented on: 11/4/2013 8:30:00 - 9:00:00

We address one of the major computational problems in the pharmaceutical and agrochemical business: the rational design of delivery vehicles for badly soluble actives. For example, in a typical scenario in a discovery/development cycle, one has only a few months, or even weeks, to find a proper formulation, driven by very costly time expenditure way down the pipeline. The chemical space for formulations is enormous, and even if one considers modern high-throughput experimentation, chances are the optimal formulation cannot be found by experimentation alone. For chemical informatics, such design also poses quite some challenges, since classical descriptor technology is almost always unitary (component specific), and therefore not suited for the non-linear binary interactions between active and delivery matrix. The more so, since in many cases one seeks or tests delivery systems that are based on some self-assembly structure, such as an emulsion, micelle or liposome. In contrast, in existing chemical informatics methods for the much simpler logP predictions one can rely on such unitary descriptors quite well, since the (iso-octanol/water) matrix in this case is known and constant over a known training set. Our strategy is completely different. We develop computational screening technologies that do take into account the structural resolution of both active and matrix, by employing a novel concept of thermodynamic fragmentation. Thereby the interactions of drug-matrix is described as a interaction between fragments, with the advantage that the fragmentation can be calibrated by (non-pharma) engineering databases. The presentation discusses in depth the motivation, the algorithm and some highlights from recent applications.

(56b) Computational Support of Solid Form Selection in Pharmaceutical Industry

Abramov, Y., Pfizer Global Research & Development
Originally presented on: 11/4/2013 9:00:00 - 9:20:00

In the pharmaceutical industry, drug polymorphism can be a critical problem, and is the subject of various regulatory considerations. One of the principal concerns is based on an effect that polymorphism may have on a drug's bioavailability due to change of its solubility and dissolution rate. Therefore, in Pharmaceutical industry, it is quite crucial to get comprehensive information on the available drug polymorphs and their relative stability and solubility.

State-of-the art computational approaches were proposed and developed to support the commercial solid state nomination. Rational solvent selection is adopted to guide polymorph screening experiments. In addition, computational methods enable an estimation of the likelihood of existence of an unknown, more stable form of the drug, as well as its potential impact on drug's solubility (and bioavailability). These computational methods involve crystal structure prediction, prediction of drug solubility and conformer distribution in organic solvents and solvent mixtures, *in silico* hydrogen bonding propensity estimation. A detailed application of these computational tools in the pharmaceutical industry is illustrated on a selective dual ALK and c-MET inhibitor crizotinib (trade name Xalkori) and on a VEGF inhibitor axitinib (trade name Inlyta).

(56c) Reaction Path and Coverage Effects Analyses of Benzene Hydrogenation On Pd(111)

Canduela, G., Gent University
Sabbe, M. K., Universiteit Gent
Joly, J. F., IFPEN
Reyniers, M. F., Ghent University
Marin, G. B., Ghent University
Originally presented on: 11/4/2013 9:20:00 - 9:40:00

1 Introduction

Benzene hydrogenation is a key reaction in various industrially important processes such as hydrotreatment of petroleum feedstocks, caprolactam and adipic acid production, and also plays an important role in environmental protective technologies. Palladium catalysts are often employed for hydrogenation reactions because of their high activity and selectivity. The (111) surface is normally chosen in fundamental studies to evaluate the catalytic performance in benzene hydrogenation [1-3] because of its abundance on supported industrial catalysts.

Micro-kinetic modeling allows to analyze catalytic surface reactions in detail, by incorporating the surface chemistry of all elementary reaction steps into a kinetic model [4]. The catalytic hydrogenation of benzene is a complex process. Even if only one single benzene adsorption site is considered, the hydrogenation of benzene can proceed via 13 different reaction paths. Performing an extended reaction path analysis based on the results of theoretical calculations allows to identify the most important reaction steps and surface intermediates. This may lead to simplification of the complex reaction network and can contribute optimizing the industrial catalytic process.

Periodic Density Functional Theory (DFT) calculations are usually performed at low to moderate coverage. However, these theoretical calculations can also be performed at higher surface coverages, enabling the modeling of coverage effects on the kinetics and thermodynamics of the adsorption and surface reaction steps. This feature is highly important for hydrogenation reactions, since in industry these are typically performed at high hydrogen pressures, and hence, at higher hydrogen coverages.

Three different techniques are employed to perform a reaction path analysis of the full reaction network of benzene hydrogenation from periodic DFT calculations at low coverage (θ_H=0.11). Additionally, the micro-kinetic model for benzene hydrogenation is also calculated at higher hydrogen coverages (θ_H = 0.44 and 0.67), in order to evaluate the coverage effects on the simulated activities by comparison with experimental data.

2 Methodology

The adsorption energies, geometries and vibrational frequencies at 0 K are calculated from Periodic DFT calculation using the Vienna Ab initio Simulation Package (VASP 4.6) [5]. This program solves the Kohn-Sham equations of DFT by using pseudopotentials with the projector augmented wave method (PAW) and periodic boundary conditions by expansions of the wave function in terms of plane-wave basis sets for which a converged cutoff energy of 400 eV has been used.

The *Nudged Elastic Band method* [6] and *Dimer* method [7] are used to locate transition states on the potential energy surface. In addition, the harmonic oscillator approach (HO) is used to numerically calculate the vibrational frequencies of all species. In combination with statistical thermodynamics, the frequencies are used to calculate the kinetic and thermodynamic parameters of every reaction, and therefore to construct the micro-kinetic model.

In the periodic DFT calculations, one hydrocarbon molecule and one hydrogen atom are adsorbed on a (3×3) Pd(111) unit cell to evaluate the kinetics at low coverage, leading to a fractional coverage of $\theta_H = 0.11$ and of $q_{HC} = 0.11$ for the hydrocarbon species. The effect of hydrogen coverage on the kinetics and thermodynamics is evaluated by increasing the number of hydrogen atoms to 4 and 6 atoms on the (3×3) unit cell, corresponding to hydrogen coverages of 0.44 and 0.67.

Three tools are implemented in an in-house FORTRAN code for the reaction path analysis: *Degree of Rate control* (DRC) [8, 9], *Sensitivity Analysis* (SA) and *Rate Of Production* (ROP). The equations for the two first are shown below. The rate of production analysis identifies the most important reactions and/or species within the reaction network by calculating the net production rates for all intermediate species occurring in the network

3 Results and discussion

The DFT results at 0 K for $\theta_H = 0.11$ indicate that the hydrogenation of benzene to cyclohexane follows the same minimum energy path for the two preferred *bridge30°* and *hollow-hcp0°* adsorption sites of benzene on Pd(111). These two networks are defined as *bridge-site* and *hollow-site* networks, respectively. The minimum energy path is the one that runs from benzene to monohydrobenzene (BH), proceeding over cyclohexa-1,3-diene (13CHD), trihydro-1,2,3-benzene (123THB), cyclohexene (CHE) and cyclohexyl (c-hexyl) to cyclohexane (CHA). For both adsorption sites, the calculated kinetics and thermodynamics confirm this path to be potentially dominant at 450 K and indicate that the first surface reaction step has the largest activation energy. Hence, the first hydrogen addition could potentially be the step with the largest control on the rate of cyclohexane formation along the dominant reaction path, as already previously suggested by Morin et al. based from BEP relationships [10]. Furthermore, simulations of the activity for both networks indicate that the *hollow-site* is more reactive than the *bridge-site*network.

For both *bridge-site* and *hollow-site* networks, ROP analysis at actual conditions and coverages (450 K, $P_{H2} = 0.1$ bar and $P_{benzene} = 0.02$ bar) agrees with the proposed dominant path from the calculated DFT rate coefficients, where the reactions along this dominant path contribute at least 90% to the net cyclohexane rate of formation. DRC and SA indicate that for the *bridge-site* network, the second hydrogen addition along the proposed dominant path is the kinetically most significant step, while for the *hollow-site*network it is the third hydrogen addition step. Clearly, inclusion of actual surface concentrations at reaction conditions is required to identify the step with the largest control along the dominant path, since the first step has the highest activation barrier.

Next, additional periodic DFT calculations were performed at higher hydrogen coverage ($\theta_H = 0.44$ and

0.67) for the *hollow-site* dominant path obtained at $\theta_H = 0.11$. For the 0.44 hydrogen coverage case, the calculated kinetics are much faster and lead to much higher simulated catalytic activities than for the $\theta_H = 0.11$ case. These activities are in the same order of magnitude as experimentally observed at similar reaction conditions, e.g. 0.064 $mmol_{Benzene}\ s^{-1}\ kg_{cat}^{-1}$ [11]. The reaction barriers calculated at a hydrogen coverage $\theta_H = 0.67$ are even lower than for $\theta_H = 0.44$. The very weak benzene adsorption calculated at $\theta_H = 0.67$ leads however to a lower activity than using the kinetics obtained at $\theta_H = 0.44$ coverage, since the simulated benzene coverages are too low. The activity using the kinetics obtained at $\theta_H = 0.67$ remains faster than using those at $\theta_H = 0.11$. 4 Conclusions

Benzene preferentially adsorbs at the *bridge30°* and *hollow-hcp0°* adsorption sites of Pd(111). At low hydrogen coverage, the dominant path of benzene to cyclohexane at these two sites proceeds through 1,3-cyclohexadiene and cyclohexene. The first hydrogen addition step is potentially the kinetically most significant step, due to its appreciable higher activation energy compared to the other elementary reaction steps of the dominant path. The reaction path analysis shows, however, that when surface concentrations are included for the simulation of activities, the second and third reaction steps along this dominant path are the ones controlling the rate of formation for the *bridge-site* and *hollow-site*network respectively. Therefore, it is required to include actual surface concentrations at reaction conditions to identify the step with the largest control on the net rate along the dominant path.

Inclusion of the effect of hydrogen coverage on the rate coefficients in the micro-kinetic model increases the rate of formation of cyclohexane, and the simulated activities using kinetics calculated at θ_H=0.44 are in the same order of magnitude as experimental observations.

Acknowledgements

The computational resources (Stevin Supercomputer Infrastructure) and services used in this work were provided by Ghent University, the Hercules Foundation and the Flemish Government – department EWI. Finally the authors thank IFPEN and the Fund for Scientific Research Flanders (FWO). References

[1] C. Morin, D. Simon, P. Sautet, J Phys Chem B, 108 (2004) 5653-5665.

[2] F. Mittendorfer, C. Thomazeau, P. Raybaud, H. Toulhoat, J Phys Chem B, 107 (2003) 12287-12295.

[3] M. Saeys, M.F. Reyniers, M. Neurock, G.B. Marin, J Phys Chem B, 109 (2005) 2064-2073.

[4] J.A. Dumesic, D.F. Rudd, L.M. Aparicio, J.E. Rekoske, A.A. Treviño, The Microkinetics of heterogeneous catalysis, American Chemical Society, Washington, D.C., 1993.

[5] G. Kresse, J. Furthmüller, Comput. Mat. Sci., 6 (1996) 15.

[6] G. Mills, H. Jonsson, G.K. Schenter, Surf Sci, 324 (1995) 305-337.

[7] G. Henkelman, H. Jonsson, Journal of Chemical Physics, 111 (1999) 7010-7022.

[8] C.T. Campbell, Journal of catalysis, 204 (2001) 520-524.

[9] C. Stegelmann, A. Andreasen, C.T. Campbell, J Am Chem Soc, 131 (2009) 8077-8082.

[10] C. Morin, D. Simon, P. Sautet, Surf Sci, 600 (2006) 1339.

[11] P. Chou, M.A. Vannice, Journal of catalysis, 107 (1987) 129-139.

(56d) Generalized Non-Linear QSPR Model for Liquid Viscosity

Yerramsetty, K., Oklahoma State University
Neely, B. J., Oklahoma State University
Gasem, K. A. M., Oklahoma State University
Originally presented on: 11/4/2013 9:40:00 - 10:00:00

Liquid viscosity (LV) is an important thermophysical property involved in many chemical processes and phenomena. Although accurate models have been developed for classes of chemicals, a need for a generalized model exists. Further, empirically developed models require experimental data, which are costly and laborious to obtain. Therefore, there exists a need for generalized models capable of providing *a priori* predictions for compounds that have no relevant experimental data or for compounds that are yet to be synthesized.

Previously, researchers at Oklahoma State University (OSU) have developed a unified framework for correlating saturation properties. This scaled-variable-reduced-coordinates (SVRC) model is based on the corresponding states theory (CST) and scaling-law behavior and in general is capable of representing saturation properties within their experimental uncertainties. The modeling approach involves the use of quantitative structure-property relationship (QSPR) methodology to generalize the model parameters of the developed SVRC model. Specifically, we used SVRC to develop the behavior model, and QSPR to generalize the SVRC model parameters. This approach, in the past, has proven to be more effective than the typical efforts to develop generalized models directly using QSPR techniques. In this work, we extended the SVRC model to correlate LV values and generalized the model parameters using structure-property modeling. A database of 5,800 data points involving 335 fluids was used in the development of this model. The predictions on the LV values were within 4% of the modeling set values. When applied to an external dataset containing 16,450 data points involving 1235 fluids, the LV model predictions were, on average, within 9% of the reported values.

(56e) New Insights On Mechanisms of Backbiting and Î²-Scission Reactions in Self-Initiated Polymerization of Methyl Acrylate

Liu, S., University of Pennsylvania
Srinivasan, S., Arkema
Grady, M., DuPont
Soroush, M., Drexel University
Rappe, A. M., University of Pennsylvania
Originally presented on: 11/4/2013 10:00:00 - 10:20:00

At high temperature, secondary reactions such as spontaneous thermal initiation, chain transfer reactions and β-scission reactions can significantly influence the polymerization process and the properties of the formed polymer [1-4]. The backbiting (intra-molecular chain transfer) and β-scission reactions have been studied extensively via various experimental techniques [5-8]. However, first-principles studies of these polymerization reactions have been hindered due to the large computational costs required for modeling polymeric systems of large size. Density functional theory (DFT) is generally regarded as a cost-effective method [9-14]. The accuracy of DFT depends on the approximation of the exchange-correlation function, and there is no obvious way to decide which density functional is the best for studying a given type of chemical reaction.

In this work, we firstly carry out a benchmark study using G4(MP2)-6X, a composite quantum chemistry method,[15] to predict the thermochemical properties of two representative 1:5 backbiting reactions. We then apply three different functionals (B3LYP, M06-2X and PBE0) and wavefunction-based method, MP2 to identify the mechanisms governing backbiting and β-scission reactions and the species generated by these reactions in high-temperature self-initiated homo-polymerization of methyl acrylate. Both hindered rotor approximations [16] and harmonic approximations are applied for the estimation of activation entropy. Four mechanisms of backbiting reactions and two types of β-scission reactions are studied. Our computational results show that the 1:5 backbiting mechanism with a six-membered ring transition state and 1:7 backbiting with an eight-membered ring transition state are energetically more favored than 1:3 backbiting and 1:9 backbiting. Two types of β-scission reactions have nearly equal activation energy and rate constants. The ^{13}C-NMR and ^{1}H-NMR spectrums are simulated and compared with experimental results in order to validate the proposed mechanisms.

References

[1] Grady, M. C.; Simonsick, W. J.; Hutchinson, R. A. *Macromol. Symp.* **2002**, *182*, 149.

[2] Quan, C. L.; Soroush, M.; Grady, M. C.; Hansen, J. E.; Simonsick, W. J. *Macromolecules* **2005**, *38*,

7619.

[3] Chiefari, J.; Jeffery, J.; Mayadunne, R. T. A.; Moad, G.; Rizzardo, E.; Thang, S. H. *Macromolecules* **1999**, *32*, 7700.

[4] Zorn, A. M.; Junkers, T.; Barner-Kowollik, C. *Macromol. Rapid Commun.* **2009**, *30*, 2028.

[5] Junkers, T.; Bennet, F.; Koo, S. P. S.; Barner-Kowollik, C. *Journal of Polymer Science Part a-Polymer Chemistry* **2008**, *46*, 3433.

[6] Yamada, B.; Oku, F.; Harada, T. *Journal of Polymer Science Part a-Polymer Chemistry* **2003**, *41*, 645.

[7] Asua, J. M.; Beuermann, S.; Buback, M.; Castignolles, P.; Charleux, B.; Gilbert, R. G.; Hutchinson, R. A.; Leiza, J. R.; Nikitin, A. N.; Vairon, J. P.; van Herk, A. M. *Macromol. Chem. Phys.* **2004**, *205*, 2151.

[8] Wang, W.; Hutchinson, R. A. *Chemical Engineering & Technology* **2010**, *33*, 1745.

[9] Degirmenci, I.; Aviyente, V.; Van Speybroeck, V.; Waroquier, M. *Macromolecules* **2009**, *42*, 3033.

[10] Van Speybroeck, V.; Van Cauter, K.; Coussens, B.; Waroquier, M. *ChemPhysChem* **2005**, *6*, 180.

[11] Furuncuoglu, T.; Ugur, I.; Degirmenci, I.; Aviyente, V. *Macromolecules* **2010**, *43*, 1823.

[12] Srinivasan, S.; Lee, M. W.; Grady, M. C.; Soroush, M.; Rappe, A. M. *J. Phys. Chem. A* **2010**, *114*, 7975.

[13] Liu, S.; Srinivasan, S.; Grady, M. C.; Soroush, M.; Rappe, A. M. *J. Phys. Chem. A* **2012**, *116*, 5337.

[14] Moghadam, N.; Liu, S.; Srinivasan, S.; Grady, M. C.; Soroush, M.; Rappe, A. M. *J. Phys. Chem. A* **2013**, *117*, 2605.

[15] B. Chan, J. Deng, L. Radom J. Chem. Theory Comput., 7, 112-120 (2011).

[16] Yu, X. R.; Pfaendtner, J.; Broadbelt, L. J. *J. Phys. Chem. A* **2008**, *112*, 6772.

(56f) A Theoretical Study of Mechanisms of Alkyl Acrylate Polymer Chain Transfer to Several Solvents

Moghadam, N., Drexel University
Liu, S., University of Pennsylvania
Srinivasan, S., Arkema
Grady, M. C., DuPont Experimental Station
Soroush, M., Drexel University
Rappe, A. M., University of Pennsylvania
Originally presented on: 11/4/2013 10:20:00 - 10:40:00

Acrylates are principal monomers in the production of coatings, adhesives, and polymers used in medical and also pharmaceutical applications, due to their transparency and resistance to breakage [1, 2]. The basic design of resins used in automobile coating has been changed because of the environmental limits on allowable volatile organic contents (VOCs) of resins [3, 4]. High-temperature (>100ºC) polymerization of alkyl acrylates allows for reducing the solvent content of the resins, while keeping the resin viscosity within a desired range [5, 6]. However, at the high temperatures secondary reactions such as spontaneous initiation, backbiting, *β*-scission, and chain transfer to monomer, polymer and solvent reactions have stronger effects on the quality of the resins [7-10].

Previous studies [11-14] showed the successful use of computational quantum chemistry to better understand self-initiation, propagation, and co-initiation reactions in polymerization of alkyl acrylates. In our previous work [15, 16], we used different functionals (B3LYP, X3LYP, M06-2X) to study several mechanisms of chain transfer to monomer (CTM) reaction in thermal polymerization of methyl, ethyl, and *n*-butyl acrylate (MA, EA, *n*-BA). The kinetics of the most likely chain transfer to polymer (CTP) mechanisms was also studied for MA, EA, and *n*-BA [17]. To the best of our knowledge, no computational study of chain transfer to solvent (CTS) reactions in thermal polymerization of alkyl acrylates has been conducted as of yet.

This paper presents a computational study of CTS reaction mechanisms in self-initiated high-temperature polymerization of MA, EA, and *n*-BA in solvents such as butanol (polar, protic), methyl ethyl ketone (MEK) (polar, aprotic), and *p*-xylene (nonpolar). Three types of hybrid functionals (B3LYP, X3LYP, and M06-2X) and three different basis sets (6-31G(d,p), 6-311G(d), and 6-311G(d,p)) have been applied to predict energy barriers and molecular geometries of reactants, products, and transition states. The sensitivity of the levels of theory to barriers and rate constants has been evaluated. The effects of live polymer chain length and the type of mono-radical that initiated the live polymer chain on the kinetics of CTS reactions have also been investigated. Solvent effects have been explored for CTS reactions using polarizable continuum model (PCM). Among chain transfer to *n*-butanol, *sec*-butanol, and *tert*-butanol reactions, the *tert*-butanol reaction has the highest energy barrier and the lowest rate constant. We have identified that the type of mono-radicals generated via self-initiation had little or no effect on the ability of the MA, EA, and *n*-BA live polymer chains to undergo CTS reactions. The energy barriers do not change significantly with the length of the live polymer chains. Lower activation energy and higher rate constant of chain transfer to *n*-butanol reaction (than those estimated for chain transfer to MEK and *p*-xylene reactions) indicate the higher ability of *n*-butanol to facilitate chain transfer.

References

[1] Drin, A. P.; Efanova, V. V.; Shut, N. I. J. Eng. Phys. Thermophys. 1994, 66, 164-171.

[2] Okor, R. S. J. Controlled Release 1990, 12, 195-200.

[3] Schultz, K.R., in *Proc. Adv. Coat. Technol. Conf.* 1992: Chicago, IL.

[4] VOC's Directive, EU Committee of the American Chamber of Commerce in Belgium, ASBL/VZw, Brussels, July 8. 1996.

[5] Campbell, J. D., F. Teymour, and M. Morbidelli, High temperature free radical polymerization, 1. Investigation of continuous styrene polymerization, Macromolecules, 2003. 36(15): p.5491-5501.

[6] Wang, W. and R. A. Hutchinson, Recent advances in the study of high-temperature free radical acrylic solution copolymerization, Macromolecular Reaction Engineering, 2008. 2(3):p. 199-214.

[7] Grady, M. C.; Simonsick, W. J.; Hutchinson, R. A.; Studies of Higher Temperature Polymerization of *n*-Butyl Methacrylate and *n*-Butyl Acrylate, *Macromol. Symp.* 2002, 182, 149-168.

[8] Rantow, F. S.; Soroush, M.; Grady, M.; Kalfas, G.; Spontaneous Polymerization and Chain Microstructure Evolution in High Temperature Solution Polymerization of *n*-Butyl Acrylate, *Polymer* 2006, 47, 1423-1435.

[9] Nikitin, A. N.; Hutchinson, R. A.; Wang, W.; Kalfas, G. A.; Richards, J. R.; Bruni, C.; Effect of Intramolecular Transfer to Polymer on Stationary Free-Radical Polymerization of Alkyl Acrylates, 5 – Consideration of Solution Polymerization up to High Temperatures, *Macromol. React. Eng.* 2010, 4, 691-706.

[10] Quan, C.; Soroush, M.; Grady, M. C.; Hansen, J. E.; Simonsick, W. J.; High-Temperature Homopolymerization of Ethyl Acrylate and *n*-Butyl Acrylate: Polymer Characterization, *Macromolecules* 2005, 38, 7619-7628.

[11] Yu, X.; Pfaendtner, J.; Broadbelt, L.J.; Ab Initio Study of Acrylate Polymerization Reactions: Methyl Methacrylate and Methyl Acrylate Propagation, J. Phys. Chem. A. 2008, 112, 6772-82.

[12] Srinivasan, S.; Lee, M.W.; Grady, M.C.; Soroush, M.; Rappe, A.M.; Computational Study of the Self-Initiation Mechanism in Thermal Polymerization of Methyl Acrylate, J. Phys. Chem. A 2009, 113, 10787-94.

[13] Srinivasan, S.; Lee, M.W.; Grady, M.C.; Soroush, M.; Rappe, A.M.; Self-Initiation Mechanism in Spontaneous Thermal Polymerization of Ethyl and n-Butyl Acrylate: A Theoretical Study, J. Phys. Chem. A 2010, 114, 7975-83.

[14] Liu, S., S. Srinivasan, M.C. Grady, M. Soroush, and A.M. Rappe, Computational Study of Cyclohexanone-Monomer Co-initiation Mechanism in Thermal Homo-polymerization of Methyl Acrylate

and Methyl Methacrylate. *Journal of Physical Chemistry A*, 2012. 116(22): p. 5337-5348.

[15] Moghadam, N.; Soroush, M.; Liu, S.; Srinivasan, S.; Rappe, A. M.; Grady, M. C.; A Theoretical Study of Mechanisms for Chain Transfer to Monomer Reactions in Alkyl Acrylates, Paper 279e, AIChE Annual Meeting 2012.

[16] Moghadam, N.; Liu, S.; Srinivasan, S.; Grady, M. C.; Soroush, M.; Rappe, A. M.; Computational Study of Chain Transfer to Monomer Reactions in High-Temperature Polymerization of Alkyl Acrylates, J. Phys. Chem. A 2013, 117, 2605-2618.

[17] Moghadam, N.; Soroush, M.; Liu, S.; Srinivasan, S.; Rappe, A. M.; Grady, M. C.; Kinetics of Mechanisms of Chain Transfer to Polymer Reactions in Alkyl Acrylates: A Theoretical Study, Paper 508e, AIChE Annual Meeting 2012.

(56g) Computer-Aided Solvent and Process Design for Carbon Capture

Qadir, A., The University of Sydney
Dyson, A., The University of Sydney
Sharma, M., The University of Sydney
Arab, M., The University of Sydney
Khalilpour, R., The University of Sydney
Chiesa, M., Laboratory of Energy and Nano-Science (LENS)
Abbas, A., The University of Sydney
Originally presented on: 11/4/2013 10:40:00 - 11:00:00

The design of novel solvents for process improvements has stemmed much interest in recent years (Bardow et al., 2010, Folić et al., 2007, Oyarzun et al., 2011). Solvents are used in many industrial processes for separation of useful products from a feed. Consequently improvements in solvent performance are sought after for a more efficient and cost effective process. The properties desired in a solvent vary from process to process and include selectivity, volatility, flammability, toxicity, viscosity, heat capacity and reaction kinetics. Often a trade-off has to be reached between the desired properties as not all desired properties may be achievable within one solvent.

In order to optimize system performance, both process and solvent parameters must be optimized. The simultaneous optimization of a process and solvent is a challenging task. Although process parameters may be continuously optimized, solvent properties are discrete and therefore pose a major hurdle in the optimization process. Complex mixed-integer nonlinear programming (MILP) methods have to be implemented in order to optimize the nonlinear process problem and choose solvent parameters from a discrete solvent property space. This problem has traditionally been circumvented by decoupling the solvent and process optimization processes and by optimizing the solvent to specific desired property targets. The Computer Aided Molecular Design- Continuous Molecular Targeting (CAMD-CoMT) framework proposed by Bardow et al is an effective way of tackling the integrated process and solvent design problem and does not require decoupling the two optimization problems. Instead, the optimization problem assumes a hypothetical solvent with continuous solvent parameters based on a physically based thermodynamic model. The perturbed-chain-statistical associating fluid theory (PC-SAFT) equation of state is used to calculate the solvent properties based on five solvent parameters.

In the current paper, we calculate optimal process parameters and determine two optimal solvents (pure and binary mixture) for a carbon capture process in contrast to the pure solvent search in presented in (Bardow et al., 2010). The process is modelled in ASPEN Plus (Aspentech, USA), while an optimization is executed in Matlab (Mathworks, USA) to calculate the vapour liquid equilibrium properties of the solvent using the PC-SAFT thermodynamic model. This simultaneous process and solvent optimization enhances the performance of the system considerably, and has good potential for application across a myriad of other process systems.

BARDOW, A., STEUR, K. & GROSS, J. 2010. Continuous-Molecular Targeting for Integrated Solvent and Process Design. *Industrial & Engineering Chemistry Research,* 49,2834-2840.

FOLIĆ, M., ADJIMAN, C. S. & PISTIKOPOULOS, E. N. 2007. Design of solvents for optimal reaction rate constants. *AIChE Journal,* 53,1240-1256.

OYARZUN, B., DARDOW, A. & GROSS, H. 2011. Integration of process and solvent design towards a novel generation of CO2 absorption capture systems. *Energy Procedia,* 4, 282-290.

(63a) Relative Free Enthalpies for Point Mutations in Two Proteins With Highly Similar Sequence But Different Folds

Hansen, N., Swiss Federal Institute of Technology
Allison, J. R.,
Hodel, F. H., Swiss Federal Institute of Technology
van Gunsteren, W. F., Swiss Federal Institute of Technology
Originally presented on: 11/4/2013 8:30:00 - 8:45:00

Understanding the capability of a protein to change to a completely different fold as a result of minor mutagenic perturbation is central to understanding how new protein structures and functions evolve [1]. Currently the structures of protein sequences at the interface between folds cannot be predicted reliably with both *ab initio* and knowledge based prediction algorithms [1]. A pair of proteins derived from Streptococcus protein G, GA95 and GB95, that have different folds despite being different by only three amino acids [2] is one striking example demonstrating that small local changes can cause large global changes and has implications for sequence-based structure-prediction methods. To understand why the amino acid changes bring about this large conformational change we use molecular dynamics simulations in explicit solvent and employ the enveloping distribution sampling methodology [3,4] to calculate the changes in free enthalpy associated with point mutations in the two different folds. The results show that point mutations at different positions in the amino acid sequence do not contribute equally to the free enthalpy difference between the native and non-native sequence-structure combinations. The individual free-enthalpy changes are almost sequence independent in the four-strand/one-helix structure, the stable form of GB95, while in the three-helix bundle structure, the stable form of GA95, an interplay between residues 20 and 45 is observed, supporting the experimental findings for this protein.

[1] P. N. Bryan and J. Orban, *Curr. Opin. Struct. Biol.* (2013) http://dx.doi.org/10.1016/j.sbi.2013.03.001

[2] P. A. Alexander, Y. He, Y. Chen, J. Orban and P. N. Bryan, *Proc. Natl. Acad. Sci. USA* **106** (2009) 21149

[3] C. D. Christ and W. F. van Gunsteren, *J. Chem. Phys.* **126** (2007) 184110

[4] N. Hansen, J. Dolenc, M. Knecht, S. Riniker and W. F. van Gunsteren, *J. Comput. Chem.* **33** (2012) 640

(63b) Activation Mechanism of c-Src Tyrosine Kinase

Shukla, D., Stanford University
Roux, B., University of Chicago
Pande, V., Stanford University
Originally presented on: 11/4/2013 8:45:00 - 9:00:00

Mechanistic understanding of the large scale conformational transformations coupled with activation of kinases is of enormous importance for treating aberrant kinase signaling, which leads to uncontrolled growth and differentiation of cancerous cells. Atomistic molecular dynamics simulations of kinases could capture these conformational transitions. However, due to the long timescales (100's of μs) associated with activation, such simulations are not feasible except on dedicated hardware. In this study, we employ an alternative computational paradigm that couples transition pathway generation techniques and Markov State Model (MSM) based massively distributed simulations for mapping the conformational landscape of c-src tyrosine kinase. MSMs of c-src tyrosine kinase capture the thermodynamics and kinetics of kinase activation for the first time, and help identify key intermediates along the activation pathway. We also predict the existence of a novel allosteric site in c-src intermediate state, which could be potentially utilized for drug design. Furthermore, kinetic model of kinase autophosphorylation built using the data obtained from atomistic simulations predict the experimentally observed induction time in the time evolution of active kinase population.

(63c) Dynamics of Torsionally Stressed Chromatin Fiber

Nam, G. M., University of California, San Diego
Arya, G., University of California, San Diego
Originally presented on: 11/4/2013 9:00:00 - 9:15:00

Torsional stresses serve critical functions for chromatin fiber to participate in biological activities such as transcription and replication of DNA. Despite the biological significance, most experimental and theoretical work so far has focused on understanding supercoiling in naked DNA, and little is known about how torsional forces are propagated, stored, and mitigated within chromatin fiber. In this work, we investigate the dynamic properties of nucleosome arrays subjected to torsional forces using the Brownian Dynamic simulations of a coarse-grained model of the chromatin fiber. By applying the twist to one end of array in stepwise manner, we monitor the conformational changes of array by measuring its vertical extension as a function of the number of turns. The obtained extension-rotations curve shows asymmetric shape with a shifted maximal extension toward negative rotations, which is in good qualitative agreement with the earlier experiments. These features imply that the nucleosome array exhibits different torsional behaviors in the two opposite directions of rotations: Negative rotations promote the linker DNA wrapping around nucleosome, promoting a one-start solenoidal structure while positive rotations cause the linker DNA to be bent in opposite direction to the handedness of nucleosome, leading to formation of a zig-zag structure. These higher-order structures are characterized by the two-angle model and their dynamical organization can be elucidated in terms of twist propagation. By means of rotational autocorrelation function for each segment along the array, we show that nonuniform twist profiles reflect the existence of geometrical barrier for twist propagation, which modulate the internal and global organization of array under torsional stresses.

(63d) Effects of DNA Methylation On DNA-Histone Interactions and Nucleosome Positions

Mathew, S., Columbia University
Wong, K. Y.,
Ortiz, V., Columbia University
Originally presented on: 11/4/2013 9:15:00 - 9:30:00

DNA occurs in CpG dinucleotides when a methyl group is added at the carbon 5 position of the C base ring, and it has an important role in regulating gene expression in higher eukaryotes. It is known that when cytosine bases are methylated, gene is silenced, but the precise role of methylated DNA in this process is still debated, and the mechanism by which methylation affects gene expression remains poorly understood. One way in which DNA methylation could regulate expression is by modifying the positions of nucleosomes, which in turn can regulate gene transcription by blocking the access of transcription factors to the gene. Here, we test this hypothesis by performing MD simulations of short DNA segments in solution and in the nucleosome, using an all atom force field. DNA conformations are characterized using the six rigid body base-pair step parameters, minor groove width, hybridization state and phosphate solvent exposure, and the changes in free-energy of the nucleosome are calculated upon cytosine methylation.

(63e) Scaling Laws and Essential Physics of DNA Oligomer Hybridization Rate Constants

Hinckley, D. M., University of Wisconsin-Madison
de Pablo, J. J., University of Wisconsin-Madison
Originally presented on: 11/4/2013 9:30:00 - 9:45:00

DNA hybridization is a fundamental process in biology and, increasingly so, in engineering. Theory and experiment have been developed to predict the scaling of hybridization rates constants for strands of 100 or more base pairs in length; however, the theory for hybridization of short oligomers remains undeveloped. In this work, we couple a recently developed version of the 3-Site-Per-Nucleotide (3SPN) DNA model with Forward Flux Sampling to examine the hybridization of short oligomers. We begin by studying the scaling of the rate constants on length for homogeneous and heterogenous sequences. We then determine the possible sites for nucleation and the mechanisms of hybridization. Lastly, we contrast the mechanisms of hybridization to those observed for DNA melting on the same sequences. These results are discussed in the context of polymer physics theory and previous simulations using simpler models.

(63f) Developing Predictive Models for Peptoids

Weiser, L., North Carolina State University
Santiso, E. E., Massachusetts Institute of Technology
Originally presented on: 11/4/2013 9:45:00 - 10:00:00

Peptoids (poly-n-substituted glycines) are peptide-like macromolecules composed of modified glycine units which have side chains attached to the nitrogen atoms [1]. Peptoids are biocompatible and resistant to protease degradation [2], and peptoid sequences can be synthesized using more than 300 commercially available amines [1], [3], making peptoids versatile and appealing compounds for combinatorial materials design. Additionally, peptide backbones don't have hydrogen bonding making developing peptoid models less challenging than developing models for their peptide counterparts.

The goal of this project is to develop predictive coarse-grained models for peptoids that can be used for *in silico* combinatorial design. This will be achieved in two stages: (1) Development of a model for the peptoid backbone that can reproduce its structural properties under different conditions, and (2) Development of models for a library of side chains that can later be used for combinatorial design. We will start by developing atomistic models using the OPLS force field with missing parameters fit to *ab initio* calculations. The models will then be fine-tuned to reproduce experimental structure data.

We will present models of peptoids with simple side chains. Atomistic simulations will be used to develop coarse-grained models. Our models will be tuned to accurately reproduce structural properties obtained by NMR measurements, and compared to available literature data. In future work, we will determine interaction potentials for other peptoid side-chains. These models will serve as starting points in developing predictive models for the combinatorial design of more complex peptoid sequences.

References:

[1] J. Seo, B. Lee, Z. R. N. P. Synthesis, N. I. P. Ducheyne, K. E. Healy, D. W. Hutmacher, D. W. Grainger, and C. J. Kirkpatrick, *Comp Biomaterials.* **2**, 53 (2011)

[2] K. T. Nam, S. a Shelby, P. H. Choi, A. B. Marciel, R. Chen, L. Tan, T. K. Chu, R. a Mesch, B.-C. Lee, M. D. Connolly, C. Kisielowski, and R. N. Zuckermann, *Nature materials.* **9**, 454 (2010)

[3] R. N. Zuckerman, S. B. H. Kent, and W. H. Moost. *J. Am. Chem. Soc.* **6,** 10646 (1992)

(63g) Structure and Dynamics of Mixtures of Ionic Liquids and Polyethers

Sharma, A., University of Notre Dame
Maginn, E. J., University of Notre Dame
McCready, M. J., University of Notre Dame

Originally presented on: 11/4/2013 10:00:00 - 10:15:00

Ionic liquids are salts that remain liquid near ambient temperatures. Characteristics such as extremely low vapor pressure and excellent thermal stability have led them to receive significant attention as potential solvents for separation processes and a plethora of other applications. One drawback of ionic liquids, however, is that they tend to be more viscous than conventional solvents. The large viscosity may be a limiting factor for the use of ionic liquids in industrial operations due to high mass transfer resistance and increased pumping and handling costs. One strategy being used to reduce the viscosity of ionic liquids is to mix them with small amounts of low viscosity additives such as polyethers or other low volatility fluids.

In order to understand the role additives have on the structure and dynamics of ionic liquids, we have carried out extensive molecular dynamics simulations on mixtures containing imidazolium-based ionic liquids and polyethers. Viscosities, self-diffusivities and rotational dynamics were computed as a function of polyether concentration and temperature. Structural properties such as radial and spatial distribution functions and end-to-end vectors were computed and used to help understand the trends in the dynamics. Comparison with available experimental data shows that the simulations are able to accurately capture both pure fluid and mixture dynamics.

(63h) Liquid-Vapor Phase Behavior of Asphaltene-Like Molecules

Kumar, V., University at Buffalo
Rane, K. S., University at Buffalo, The State University of New York
Wierzchowski, S., Shell Global Solutions
Errington, J. R., University at Buffalo, The State University of New York
Originally presented on: 11/4/2013 10:15:00 - 10:30:00

Asphaltenes have a significant effect on the thermodynamic and transport properties of crude oil. These molecules are known to clog pipes as well as clog the pores within rocks in the oil-field, thereby impeding oil extraction. The identification and design of solutions to mitigate the impact of asphaltenes would benefit from a better understanding of their thermophysical properties, including their phase behavior. In this presentation, we describe our recent efforts aimed at using molecular modeling to construct liquid-vapor phase diagrams for asphaltene-like molecules. A collection of histogram-based Monte Carlo (MC) techniques are used to deduce saturation properties. We first use direct grand canonical simulations to acquire coexistence properties at relatively high temperatures. These calculations are followed by temperature expanded ensemble simulations, which provide a means to trace saturation curves to relatively low temperature. Reservoir grand canonical MC, hybrid MC, and growth expanded ensemble strategies are employed to facilitate sampling the configuration space of these complex molecules. We first provide results for the phase behavior of a collection of polycyclic aromatic molecules. Such structures form the central unit of asphaltenes, and therefore represent a natural starting point for understanding the behavior of these molecules. Next, we probe the extent to which nitrogen- and sulfur-containing chemistries influence the phase behavior of polycyclic aromatics. Finally, we examine the phase behavior of realistic asphaltenes by decorating polycyclic aromatic molecules with tethered alkane chains.

(63i) Replica Exchange Dynamic Molecular Simulations and Studies of Novel Cyclic and Linear Amphiphilic Homopolymer

Liu, L., Tulane University

Originally presented on: 11/4/2013 10:30:00 - 10:45:00

Replica Exchange Dynamic Molecular simulations of Cyclic and Linear Amphiphilic Homopolymer

Lixin Liu, Henry S. Ashbaugh*, Steve W. Rick*, Scott Grayson*

Department of Chemical and Biomolecular Engineering, Tulane University, New Orleans, LA 70118

Abstract

Polymer-based micelles had been raised particular interests in application as drug delivery material because their spontaneously forming core-shell structure with hydrophobic core encapsulating solvent-compatible particles as long as hydrophilic corona imparting water-solubility upon the aggregation. In our research work, Molecular Dynamic (MD) simulations based on a novel Amphiphilic Homopolymer c-PHS-Amph first reported by Scott Grayson's lab were performed to investigate polymer's unique topologies as well as its resulting prominent solvent-dependent character, which would progressively shed a in-depth interpretation to the molecular driving force leading amphiphilic behavior. Since Macromolecule requiring relative large scale computational work, we have cooperated with Steve Rick's group to employ Replica Exchange Dynamic Simulation (REDs) to improve sampling space and determine the global polymer conformation. Through applying REDs technique, exchange among replicas split in large potential barriers occurred efficiently and successfully deliver linear and cyclic polymer's characteristic structure corresponding with different solvents' environment. Proximal radial distribution of solvent around hydrophobic arms and hydrophilic arms was performed to validate the thermodynamic signatures as well as used to reproduce the density profile of interacting solvents.

(123a) Model Development for the Gas-to-Liquid Process Based On Molecular Simulation and Experimental Measurements of Physical Properties

Economou, I. G., National Center for Scientific Research "Demokritos"
Makrodimitri, Z. A., National Center for Scientific Research "Demokritos"
Froba, A. P.,
Koller, T., 2University of Erlangen-Nuremberg
Heller, A.,
Rausch, M.,
Fleys, M.,
Bos, R.,
Unruh, D. J. M.,

Originally presented on: 11/4/2013 12:30:00 - 12:55:00

The Gas-to-Liquid (GTL) process is a highly promising technology for the production of high value fuels with a broad range of applications. Recently, Shell launched the biggest GTL plant worldwide in Qatar which in full capacity will be processing 140,000 bpd. GTL is based on Fischer-Tropsch synthesis where carbon monoxide hydrogenation takes place at elevated temperatures and pressures. For optimum process design, various physical properties are required. Some of the most critical properties include the diffusion coefficient of hydrogen (H_2), carbon monoxide (CO) and water (H_2O) in *n*-alkanes for the evaluation of mass transport process.

In this work, Molecular Dynamics (MD) simulations were performed to predict the self-diffusion coefficient of the three gases in various *n*-alkanes, from *n*-C_8 up to *n*-C_{96} and their mixtures, at temperature and pressure conditions similar to the actual GTL process. Realistic atomistic force-fields were used to model gas and *n*-alkane intra- and intermolecular interactions. The force-fields were initially validated for the prediction of pure component thermodynamic properties. Subsequently, they were used for the prediction of gas diffusivity in pure *n*-alkanes and in binary and multicomponent *n*-alkane mixtures. In the latter case, the mixture compositions were chosen so that they represent closely the actual GTL mixtures. MD calculations are in excellent agreement with limited experimental data for the diffusion of gases in medium size *n*-alkanes. In addition, the Maxwell-Stefan diffusion coefficient of H_2 and CO in *n*-C_{12}, *n*-C_{20} and *n*-C_{28} was calculated for different solute concentrations and used subsequently for the calculation of the Fick diffusion coefficient.

A high pressure Dynamic Light Scattering (DLC) unit was built in the University of Erlangen-Nuremberg to measure the diffusion coefficient of the gases in *n*-alkanes. Experimental data for the diffusion coefficient of the three gases in *n*-C_{12} and *n*-C_{28} in a wide temperature range agree remarkably well with the MD predictions.

MD simulations are proven to be very accurate for property predictions but they are computationally demanding and thus cannot be used routinely for engineering calculations needed for process modeling. Consequently, an empirical phenomenological model was developed that correlates the diffusion coefficient of a gas with respect to other physical properties, such as the *n*-alkane viscosity. The latter was also calculated based on MD simulations using the same force-field. Calculations based on the empirical model are, on the average, within 10 % from MD simulations.

Finally, work is under way to predict the diffusion coefficient of gases in 1-alcohols which may be present in small quantities in the GTL process.

(123b) Simulation Studies of the Effect of Reactant Molecular Architecture On the Mechanical Properties of Thermosetting Polymers

Rigby, D., Materials Design, Inc
Saxe, P. W., Materials Design, Inc
Freeman, C. M., Materials Design, Inc.
Originally presented on: 11/4/2013 12:55:00 - 13:20:00

Simulation Studies of the Effect of Reactant Molecular Architecture on the Mechanical Properties of Thermosetting Polymers

Modifying the properties of thermosetting polymers used in composite systems by manipulating the chemistry and molecular architecture of resin components is of considerable interest in advanced engineering applications. Thus for example, there is a need to understand how chemistry and thermoset topology affect fundamental material properties such as small strain elastic behavior, material toughness, thermal and electrical properties, and aspects of behaviour such as adhesion and the tendency to undergo microcrack formation under conditions of temperature cycling and exposure to wet or humid environments.

In recent simulation work on polymer blends containing polystyrene and poly(2,6-dimethyl-1,4-phenylene oxide), it was shown that atomistic simulation using accurate force fields can be used effectively to predict relatively small changes in tensile moduli of the material as a function of blend composition when the results are analyzed using the Hill-Walpole bounds analysis approach applied to polymers by Suter and Eichinger [1-4]. The present work accordingly extends the earlier investigation by applying the approach to models of thermoset polymers used as composite matrix material.

After briefly summarizing the earlier studies [4], we present a description of the thermoset building approach and examine its ability to create defect-free densely crosslinked systems with high degree of conversion. This is followed by a study of epoxy thermosets in which the effect of varying reactant architecture using resins with chemical functionalities of 2, 3 and 4 (so-called RA_4+$R'B_2$, RA_4+$R'B_3$ and RA_4+$R'B_4$ polymerization types) is studied by calculating elastic constants and analyzing using the Hill-Walpole approach. The presentation will conclude with a discussion of results from ongoing studies of mechanical properties in which the epoxy-based materials are replaced by higher performance polyimide-based resin.

References:

1. Hill, R. J. Mech. Phys. Solids **13**, 213 (1965).
2. Walpole, L.J. J. Mech. Phys. Solids **14**, 151 (1966).
3. Suter, U.W. and Eichinger, B.E., Polymer **43**, 575 (2002).
4. Rigby, D., Saxe, P.W. and Freeman, C.M., 2012 AIChE Annual Meeting, Pittsburgh, PA.; manuscript in preparation.

(123c) Thermorheological Behavior of Model Asphalt

Masoori, M., University of Rhode Island
Greenfield, M. L., University of Rhode Island
Originally presented on: 11/4/2013 13:20:00 - 13:45:00

Molecular simulations of asphalts provide a method to understand how the structure and modes of molecular motions contribute to the Asphalt viscoelastic response to applied strain or stress. Results from equilibrium molecular dynamics simulations of AAA-1 model asphalt using two different models (prior AAA-1 model and new AAA-1 model) have been investigated by calculating storage and loss moduli from stress relaxation modulus G(t) data from molecular simulation. Storage and loss moduli have been exploited to calculate complex modulus and phase angle, which are more typical for civil engineers. All models in different temperature show similar phase angle at very high frequency, with the slope of the lower frequency side of the peak increasing with temperature. The slope of complex modulus with frequency shows an opposite temperature dependence, with a faster rise with frequency at higher temperature. To have better understanding of rheological behavior of new AAA-1 model asphalt, van Gurp-Palmen plot of complex modulus vs. phase angle has been plotted for this model asphalt which shows deviations from thermorheological simplicity over this temperature range. The prior AAA-1 model system, composed of smaller molecules, shows a quantitatively lower complex modulus.

(123d) Molecular Simulation of Aqueous and Non-Aqueous Electrolyte Solutions

Reiser, S.,
Deublein, S., University of Kaiserslautern
Vrabec, J., University of Paderborn
Hasse, H., University of Kaiserslautern
Horsch, M.,
Originally presented on: 11/4/2013 13:45:00 - 14:10:00

Electrolyte solutions play an important role in many industrial processes. In the present work, we apply molecular modeling and simulation on the basis of classical force fields with explicit solvent models to determine the thermodynamic properties of aqueous and non-aqueous electrolyte solutions. The focus of this work is on alkali cations and anions from the halide group. The ions are modeled as one Lennard-Jones (LJ) site with a point charge of ± 1e in its center of mass. For aqueous solutions, SPC/E water is used which is of the LJ + partial charge type. The force fields of the non-aqueous solvents, namely methanol and ethanol, are taken from previous work of our group. These models consist of two and three LJ sites, respectively, and three partial charges to model both polarity and hydrogen bonding. The simulations were performed with the molecular simulation program *ms*2. The long range interactions of the charges were considered by Ewald summation.

A comprehensive set of new force fields for the ions of the groups specified above is developed. They are optimized for describing basic thermodynamic data of aqueous solutions. The size parameter is adjusted to the reduced liquid solution density at 293.15 K, whereas the energy parameter (which does not significantly influence the density) is fitted to self-diffusion coefficient data at 298.15 K. A global fitting approach is used. The new models perform well over a wide range of ionic strength for all combinations of cations with anions. They predict well structural properties like the hydration number and transport properties like the electric conductivity in aqueous electrolyte solutions. The models are also studied with respect to the temperature dependence of the reduced density of the electrolyte solutions up to 333.15 K. The predictions from molecular simulation are in good agreement with experimental data, part of which is measured in the present work.

Basic thermodynamic properties of non-aqueous electrolyte solutions with the solvents methanol and ethanol are predicted by molecular simulation for all cation/anion combinations from the list above. The Lorentz-Berthelot combining rule is used for modeling the unlike LJ interactions between the ions and the solvent molecules. No adjustments to data for electrolytes in non-aqueous solvents are made. The predictions of the reduced liquid solution density from molecular simulation are in good agreement with experimental data, part of which is measured in the present work. Furthermore, structural properties like the radial distribution function of the solvent molecules around the ions and transport properties like the self-diffusion coefficient and the electric conductivity in non-aqueous electrolyte solutions are investigated.

(123e) Ligand Binding Free Energy Predictions Based On Single Steered Molecular Dynamics for Hydrophobic and Hydrogen Bonding Governed Interactions With An Î±-Helix

Marzinek, J., Unilever R&D Colworth Science Park
Zhao, Y., China Agricultural University
Lian, G., Unilever R&D Colworth Science Park
Mantalaris, A., Imperial College London
Pistikopoulos, E. N., Imperial College London
Bond, P.,
Noro, M.,
Originally presented on: 11/4/2013 14:10:00 - 14:35:00

Free energy is one of the most desirable and important thermodynamic properties in simulations and experiments of biological systems.[1,2] Recently, an increasing number of studies have been reported in the literature on binding free energy (ΔG) calculations of small molecules to proteins within a molecular dynamics (MD) simulation framework, often with accompanying experimental validation. However, high accuracy predictions of this thermodynamic property, in simulation systems that include explicit solvent, are computationally very expensive. Previously, we proposed a linear correlation (at given temperature and pulling rate) between the maximum pulling force (F_{max}) obtained from a single steered MD (SMD) simulation and ΔG calculated via umbrella sampling (US).[3] We reported an application of this method to epigallocatechin-3-gallate (EGCG) binding keratin filaments, the outermost skin layer protein. In the present study, we extend this to investigate the binding of four additional ligands characterized by different octanol/water partition coefficient values (ranging from -2 to 2) to the α-helical part of keratin. Together with our previous work 35 μs of simulation sampling in explicit solvent has been generated. Validation of the simulation results is performed via isothermal titration calorimetry (ITC) for three of those ligands. The calculated binding ΔG's for these small molecules are in excellent agreement with the experimental data. We also see the same relation between $F_{max}$from SMD and ΔG for each of these molecules. Hence, our results provides a means to precisely predict the binding ΔG of small molecules with various physiochemical properties to the most common protein secondary structural element via hydrogen bonds and/or hydrophobic interactions based on a single and reproducible short pulling simulation (hundreds of picoseconds) instead of time consuming US calculations (hundreds of nanoseconds).

References

1. Beveridge, D. L.; Dicapua, F. M. Free-Energy Via Molecular Simulation - Applications to Chemical and Biomolecular Systems. *Annu Rev Biophys Biophys Chem.* **1989,** *18*, 431-492.

2. Deng, Y. Q.; Roux, B. Computations of Standard Binding Free Energies with Molecular Dynamics Simulations. *J. Phys. Chem. B.* **2009,** *113*(8), 2234-2246.

3. Torrie GM, Valleau JP. Non-Physical Sampling Distributions in Monte-Carlo Free-Energy Estimation - Umbrella Sampling. Journal of Computational Physics. **1977**; 23(2):187-199.

(123f) A Molecular Dynamics Simulation Approach Towards Designing of Drug Formulations: Case Study of Anti-Cancer Drug (Taxol)

Purohit, A.,
Dutta, R. C., Tata Research Development and Design centre
Rai, B., Tata Research Development and Design centre
Originally presented on: 11/4/2013 14:35:00 - 15:00:00

An effective drug formulation shall contain the correct amount of drug in the right form and maintain its chemical and biological integrity for delivery at or over the proper time, at the proper rate, and at the desired location. Most of the new drugs under development fall into the category of Class IV compounds – low solubility and permeability. Poor solubility leads to significant hurdles in the oral absorption and bioavailability of the drug candidate by decreasing its dissolution rate and membrane permeation. A well known example of a Class IV drug is Paclitaxel (Taxol) which was first found as a natural product from the Western yew tree, *Taxus brevifolia*. It has been recognized as one of the most promising antineoplastic drugs for treatment of a wide spectrum of cancers such as breast, ovarian, colon, and lung cancer as well as melanoma and lymphoma.

The current commercial formulation of Taxol contains Cremophor-EL as solvent, which often causes serious side effects like hypersensitivity reactions, nephrotoxicity, and neurotoxicity. The search for an alternative solvent is therefore imperative. Considering the cost associated with the trial and error experimentation to arrive at possible solvent combinations, it will be beneficial to develop a molecular simulation based screening protocol to identify promising formulations for further experimental testing. On one hand, they can reduce trial and error experimentation; on the other hand, they provide insightful information at molecular as well as micro length scales which are often not extracted from experiments. The interactions computed at molecular levels can be gainfully applied to find the optimum combination for a given drug.

In this current work, a total of 8 solvent combinations were studied using molecular dynamics (MD) simulations comprising of ethanol, water, PEG 400, glycerol and Tween 80 in binary and ternary ratio (1:1 and 1:1:1). The relative solubility and relative diffusivity of Taxol in these solvent combinations is studied to find the best solvent combination. Based on solubility and diffusivity values Ethanol-Water and Ethanol-Water-Tween 80 were found to be the best solvent combinations for Taxol. The diffusivity of Taxol in Triacetin solvent, which is used for commercial formulation, was also computed and compared. The simulated properties were compared with the experimental solubility of Taxol reported in the literature.

(168a) Free-Energy Perturbation: Application to Chemical Reactions At Metal/Water Interfaces

Faheem, M., University of South Carolina
Heyden, A., University of South Carolina
Originally presented on: 11/4/2013 15:15:00 - 15:35:00

Computational investigations of structural and energetic changes in chemical reactions at metal/water interfaces pose a unique challenge of properly accounting for the effect of liquid-phase environment on free energies of reactants, products and transition states while keeping the computational model simple enough to be computationally affordable. A practical solution to this challenge is a hybrid QM/MM approach. By treating only the reaction zone and its immediate environment quantum mechanically (QM) for higher accuracy and the remainder of the metal surface and the bulk of the liquid water at the classical, molecular mechanical (MM) level of theory, a reasonably accurate energetic description of the complex metal/water system can be obtained while simultaneously realizing a computational speedup of multiple orders of magnitude.

In this talk, we present a novel solvation scheme that combines planewave density functional theory (DFT), periodic electrostatic embedded cluster (PEEC) calculations with Gaussian-type orbitals, and classical molecular dynamics (MD) calculations with the QM/MM free-energy perturbation (FEP) methodology to enable an accurate determination of free-energies of reaction and activation free-energy barriers at periodic metal/water interfaces. Optimization of local conformations (i.e., reactants, products, and transition states) are performed using an iterative approach with sequential MD sampling and QM optimization steps to reduce the number of QM calculations and the amount of MD sampling. Furthermore, we use a fixed-size, finite ensemble of MM conformations to allow a precise evaluation of the QM potential of mean force (PMF) and its gradient defined within this ensemble. As a result, we circumvent the challenges associated with statistical averaging during MD sampling and speed up the convergence of the optimization process. Application to C-C cleavage in double-dehydrogenated ethylene glycol on Pt (111) surface will be discussed as model reaction.

(168b) A Combined Density Functional Theory and Monte Carlo Approach for Quantifying Catalytic Energies in a Water Environment

Bollmann, L., Clemson University
Getman, R. B., Clemson University
Originally presented on: 11/4/2013 15:35:00 - 15:55:00

One of the biggest challenges in computational catalysis is modeling activity under realistic reaction environments. The main way that the reaction environment influences activity is by "dosing" the catalyst with atoms and molecules, which become chemically and/or physically adsorbed to the surface. These atoms and molecules alter binding and reaction energies through at least two effects: 1) molecules that form strong chemical bonds with the catalyst can accumulate on the catalyst surface, inducing coverage effects, and 2) molecules in condensed phase reaction environments can interact physically with adsorbates, inducing solvation effects. Several computational methods have been developed to treat these phenomena. For example, first-principles thermodynamics (FPT) is used to predict catalyst compositions under real reaction conditions, and a number of solvation models have been used to predict the energies of interaction between static fluid phases and solid catalyst environments. In this work, we extend these methods to more realistic, dynamic fluid environments. Specifically, we demonstrate a method to include entropic contributions. This method takes fluidity in the reaction environment and its influence on the free energies of relevant reactants, intermediates, and products into account. We use a combination of density functional theory (DFT) and Monte Carlo to quantify the different contributions to the overall free energies and make conclusions about how they influence catalytic activity. Our present focus is on reduction of nitrate over nanoparticulate Au in water. We chose Au because it is relatively non-reactive, so we can largely neglect coverage effects and instead focus on interactions between the reaction environment and the catalyst environment. Catalysts are modeled using 13-atom gold icosahedral clusters, which are computationally efficient yet comprise a number of typical adsorption sites for typical metal catalysts. We find that the water environment has a noticeable influence on reaction energetics. Ongoing work involves expanding these methods to more reactive transition metals and more realistic supported catalyst models.

(168c) Electric Field-Induced Reactions in Fuel Cells From First Principles

Che, F., Washington State University
McEwen, J. S., Washington State University
Originally presented on: 11/4/2013 15:55:00 - 16:15:00

Solid oxide fuel cells will no doubt find more widespread use in the new millennium in the conversion of chemical to electrical energy offering, as they do, very high efficiencies and a unique scalability in electricity-generation applications. To control the ongoing reactions in such a device one can modify the selectivity of the catalyst itself, which oxidizes the fuel at the anode and dissociates O_2 at the cathode. The selectivity of the catalyst in a solid oxide fuel cell will be altered by the presence of a huge electric field of the order of 1 Volts per Angstroms, which occurs near the electrode-electrolyte interface [1]. Given this background, it is surprising that the role of the electric field is often ignored in the optimization of such devices. *Ab initio* quantum chemistry methods coupled with detailed kinetic models allow the prediction of such field-induced effects. Here we report density-functional theory calculations of the dehydrogenation of methyl species on the Ni(111) surface with and without an electric field. The structures, adsorption energies, and reaction barriers for all methyl species on the Ni(111) surfaces are identified. Our results show that the presence of an electric field does not affect the structures nor does it affect the favorable adsorption sites of the adsorbed species, but causes the adsorption energies of the CH_x species and the dissociation energy barrier of a CH molecule to be significantly altered by about 0.2 eV. We expect that such field-induced effects to be even more important on a stepped surface, since the local electric field will be greatly enhanced at the step [2]. We compare these results to when a non-reactive metal (such Au) is added to the Ni catalyst with regard to the suppression of coking [3].

[1] E.M. Stuve, Ionization of water in interfacial electric fields: An electrochemical view, Chem Phys Lett, 519-520 (2012) 1-17.

[2] Y. Suchorski, N. Ernst, W.A. Schmidt, V.K. Medvedev, H.J. Kreuzer, R.L.C. Wang, Field desorption and field evaporation of metals, Prog. Surf. Sci., 53 (1996) 135.

[3] H.S. Bengaard, J.K. Nørskov, J. Sehested, B.S. Clausen, L.P. Nielsen, A.M. Molenbroek, J.R. Rostrup-Nielsen, Steam Reforming and Graphite Formation on Ni Catalysts, Journal of Catalysis, 209 (2002).

(168d) Parameterization of An Adsorbate-Adsorbate Interaction Model for Catalytic Trend Studies

Khan, T., Stanford University
Falsig, H., Technical University of Denmark
Guo, W., UNIVERSITY OF DELAWARE
Wang, S., Technical University of Denmark. Denmark
Dahl, S.,
Bligaard, T., SLAC National Accelerator Laboratory
Originally presented on: 11/4/2013 16:15:00 - 16:35:00

The lateral interaction between atoms and molecules adsorbed on the surface of a transition metal affects both the adsorption energies of the surface species and also the transition state energies for surface reactions. A simple model for obtaining the effect of such interactions on adsorption energies is proposed, where the adsorption energy varies linearly with the coverage of species at the surface. The model is parameterized using the DFT calculated adsorption energies at different coverages for a number of atoms and molecules on the most stable facet of different transition metals.

The number of parameters in a full description of the interactions varies as $N_{surface} \times N_{sp} (N_{sp} + 1)/2$, where N_{sp} is the number of adsorbed species and $N_{surface}$ is the number of surfaces we are interested in investigating. In order to reduce the number of independent parameters we discuss 3 submodels; i) Model Rh : based on the interaction parameters of Rh{111}, ii) Model GM : the cross-interaction parameters are obtained from the parameters of the interaction between the same adsorbates, iii) Model scaled : the interaction parameters are obtained from linear scaling to the adsorption energy at low coverage. We then discuss the limitations of the proposed models, and finally for the direct decomposition of NO, we show what effect the inclusion of a description of the interactions in the kinetic model, has on the catalytic activity.

(168e) Impact of Adsorbates On the Photochemistry of Metal Nanoclusters

Avanesian, T., University of California, Riverside
Kale, M., University of California, Riverside
Christopher, P., University of California, Riverside
Originally presented on: 11/4/2013 16:35:00 - 16:55:00

The possibility of photon-induced selective bond-activation of adsorbates on metal surfaces has sparked an interest in utilizing metal catalysts to drive photochemical reactions in excited states, which cannot be performed using thermal energy only. Ultrafast lasers have been used for many years to excite adsorbate covered single crystal metal surfaces and execute photo-catalytic reactions in excited states. The low efficiencies and high laser intensities required to drive these chemistries present significant barriers that must be overcome. Recent experimental work in our group has shown that spatial confinement of catalytic active sites in metal nanoparticles results in fundamentally different photo-catalytic behavior compared to extended metal surfaces. It has been demonstrated that Pt nanoparticles, diameter < 5 nm, and plasmonic Ag nanoparticles, diameter < 100 nm, are able to efficiently drive photo-catalytic reactions through resonant energy transfer processes. The goal of this study is to explain the mechanism of these nanoparticle mediated photochemical processes.

It is well known that the addition of adsorbates to metal surfaces induces electronic structure changes in the underlying metal and introduces adsorbate-derived states near the Fermi level. We have utilized density functional theory calculations to examine adsorbate induced changes in electronic structure of Pt surfaces upon addition of molecular (CO, NO) and atomic (O, C, N) adsorbates. The calculated densities of states were used as inputs into a model that describes the electron energy distribution induced by photon excitations of metal surfaces. In addition, an extended two-temperature model (ETTM) is used to describe the temporal evolution of non-equilibrium electron distributions in photo-excited metal nanoparticles. Our theoretical predictions accurately capture recent wavelength dependent photocatalytic measurements on Pt nanoparticles performed in our group and allow us to postulate general mechanisms regarding the effect of molecular and atomic adsorbates on metal nanocluster photocatalysis. The results of our study provide essential insights into the possibility of state-selected bond activation of adsorbates on metal nanoparticles under low-intensity visible light illumination.

(168f) First-Principles Modeling of Hot Electron Assisted Chemical Bond Activation On Metal Surfaces

Xin, H., University of Michigan
Linic, S., University of Michigan
Originally presented on: 11/4/2013 16:55:00 - 17:15:00

Chemical reactions on metal surfaces are typically driven by phonons, i.e., vibrational excitation of adsorbed molecules by thermal energies. Interestingly, it is well-established in the field of femtochemistry that energetic electrons, generated from optical excitation of metal single crystals with ultrashort laser pulses or other sources, can also be exploited to drive surface chemical reactions, often through novel reaction channels not accessible in thermal reactions. Recently, we showed that plasmonic nanostructures of coinage metals effectively couple thermal and low-intensity photonic stimuli to drive chemical transformations. Intensity- and wavelength-dependence of steady-state reaction rates together with kinetic isotope experiments suggest that this process is electron-driven in nature, attributed to the excitation of localized surface plasmon resonance (LSPR).

In this talk we will discuss our approach to study direct photocatalytic reactions on plasmonic nanostructures from first-principles. Photocatalytic ethylene epoxidation (catalyzed by silver (Ag) cubes) was used as the test reaction, where the O_2 dissociation is the rate-limiting step. We present a model that allows us to evaluate within a non-adiabatic dynamics the coupling of vibrational excitations of adsorbed O_2* and energetic electrons from the Landau damping of LSPR while the adsorbate-surface interaction is described by ab-initio potential energy surfaces (PESs). We used DFT calculations for the PES of O_2 dissociation on Ag(100) in its neutral state, while ΔSCF-DFT calculations were used to for the PES of the TNI state. The probability that an energetic electron with a particular energy scatters into the 2pi* anti-bonding orbital of O_2 gaining a given amount of vibrational energy is computed using the Green's function with the Newns-Anderson-type Hamiltonian. This Hamiltonian includes a non-adiabatic coupling of the energetic electron to adsorbate reaction coordinate. We substantiate that important aspects of the reaction dynamics are well captured by the model. In addition, we will also discuss electronic factors governing vibrational excitations of adsorbed species on metal surfaces and provides the rationale for manipulation of metal surfaces for electron-assisted surface reactions.

1. P. Christopher, **H. Xin**, S. Linic, Nature Chem., 3, 467, (2011)
2. P. Christopher, **H. Xin**, A. Marimuthu, S. Linic, Nature Mater. 11, 1044, (2012)
3. S. Linic, P. Christopher, **H. Xin**, A. Marimuthu, Acc. Chem. Res., accepted (2013)

(168g) Ab Initio Modeling of Direct Interaction of Surface Plasmons With Adsorbates On Metal Nanostructures

Morabito, M., University of Michigan
Xin, H., University of Michigan
Linic, S., University of Michigan
Originally presented on: 11/4/2013 17:15:00 - 17:35:00

We recently reported on a new family of photo-catalysts that can efficiently couple thermal energy and photonic stimuli to drive photo-chemical transformations with unprecedentedly high efficiencies.[1,2,3] These materials are plasmonic nanostructures of coinage metals which respond to low intensity visible light through a formation of a collective, oscillating electrons.

We will discuss our approach to study these reactions from first principles. We have employed Langevin dynamics simulations to investigate the activation of adsorbed diatomic molecules over plasmonic metal nanostructures.[4] Using these simulations we calculated molecular dissociation probabilities and reaction rates at different light intensities. These stochastic simulations combine Newton's equations of motion with electronic drag from the surface. The electronic friction is caused by partial charge transfer between substrate and adsorbed molecule due to changes in the electronic structure induced by direct interactions of adsorbates with low energy electron in plasmons. We calculated electronic friction from the adiabatic electronic wavefunctions from Density Functional Theory, using a quantum model to obtain an electron distribution in the surface plasmon, a collection of low-energy excited electrons.[5][6] We compare this approach to desorption induced by (multiple) electronic transitions (DIET/DIMET) model, another mechanism proposed for this type of plasmonic photocatalytic enhancement.

[1] P. Christopher, H. Xin, M. Andiappan and S. Linic. Nat. Mat. **11**, 1044 2012

[2] P. Christopher, H. Xin, S. Linic. Nat. Chem. **3**, 467 2011

[3] S. Linic, P. Christopher, D. B. Ingram. Nat. Mat. **10**, 911 2011

[4] T. Olsen and J. Schiøtz. J. Chem. Phys. **133**, 034115 2010

[5] V. Krishna and J. C. Tully. J. Chem. Phys. **125**, 054706 2006

[6] L. Genzel, T. P. Martin, U. Kreibig. Z. Physik B. **21**, 339 1975

(179a) Molecular Modeling of Fluid Phase Equilibria for Ternary Mixtures With Two Or Three Liquid Phases

Cortes-Morales, A. D., University of Minnesota
Harwood, D.,
Bai, P., University of Minnesota
Michalis, V., The Petroleum Institute
Peters, C. J., Petroleum Institute
Siepmann, J. I., University of Minnesota
Originally presented on: 11/4/2013 15:15:00 - 15:40:00

Multi-component mixtures that exhibit a liquid–liquid miscibility gap are of utmost importance for the oil and gas industries because of the prevalence of hydrocarbons and water in many processes. Given the complexity of such mixtures, it is not surprising that equation-of-state-based approaches perform best for the correlation of experimental measurements but are less successful for the a priori prediction of miscibility gaps. Molecular simulations of these multi-component mixtures with liquid–liquid miscibility gap are hampered by sampling difficulties (multiple phases must be treated explicitly, transfers of molecules between two liquid phases are much less likely to succeed, and spatial distribution of compounds in liquid phases is heterogeneous on the molecular scale) and by the lack of transferable force fields. Here we present configurational-bias Monte Carlo simulations in the Gibbs ensemble to investigate the vapor-liquid-liquid and vapor-liquid-liquid-liquid equilibria for binary and ternary mixtures containing water, n-decane, and 2-butoxyethanol. The predicted phase diagram is compared to experimental data [1] and the liquid-phase microheterogeneities obtained from analysis of simulation trajectories are discussed.

[1] S. Negahban, G.P. Willhite, S.M. Walas, and M.J. Michnick, “Three-liquid-phase equilibria of ternary and quaternary mixtures, water/n-decane/2-butyloxyethanol and water/n-octane/1-propanol/sodium chloride – Experimental measurements and their correlation with the UNIQUAC model,” Fluid Phase Equil. 32, 49-61 (1986).

(179b) Application of COSMO-SAC and Molecular Simulation to An Industrially Important Pentenary System

Hsieh, C. M., University of Paderborn
Merker, T., University of Kaiserslautern
Lin, S. T., National Taiwan University
Hasse, H., University of Kaiserslautern
Vrabec, J., University of Paderborn
Originally presented on: 11/4/2013 15:40:00 - 16:05:00

The oxidation of cyclohexane to cyclohexanol and cyclohexanone is an important industrial reaction and a key step in the nylon production chain. A novel route is to use supercritical carbon dioxide to expand the reactive liquid, which enhances the mobility of both the reactants and the products, to obtain a higher conversion rate and a better yield. However, the available experimental phase equilibrium data base for the design of carbon dioxide expanded oxidation processes is very narrow, especially for high pressure multicomponent systems. In this work, experimental measurements were carried out to broaden this data base, however, the number of independent variables is too large to fully explore the relevant range of states experimentally. Therefore, it is important to study the descriptive and predictive capabilities of thermodynamic models. In this presentation, three theoretical approaches, i.e. molecular simulation, the Peng-Robinson equation of state (PR EOS) and COSMO-SAC, were investigated in a comprehensive study with respect to fluid phase coexistence data of an industrially important pentenary mixture. Overall, the PR EOS provides the poorest performance, while both molecular simulation and COSMO-SAC yield good and similar results. Both molecular simulation and COSMO-SAC are capable to predict multicomponent VLE with an excellent accuracy, if one state-independent binary parameter is introduced. Moreover, it was found that improvements are necessary, especially when no experimental binary data are available. In a strictly predictive mode, deviations of up to 50% in terms of the vapor pressure or the Henry's law constant were encountered for some systems. The presented data and models can be used for the optimization of the reaction conditions for the oxidation of cyclohexane in carbon dioxide expanded liquids.

(179c) Understanding Latex Binder/Water Interfaces Using Atomistic Simulaitons

Yuan, F.,
Larson, R. G., University of Michigan
Li, Z.,
Milner, S., Pennsylvania State University
Fichthorn, K., The Pennsylvania State University
Originally presented on: 11/4/2013 16:05:00 - 16:30:00

The stability and rheology of waterborne coatings are results of the interactions of the binders, rheology modifiers as well as other additives such as surfactants. To describe the interfacial properties of binder/water interface, a slab of random copolymer of butyl acrylate and methyl methacrylate in water was simulated using atomistic molecular dynamics simulations. Density profiles of polymer and that of water were calculated to fitted in to the hyperbolic tangent function and the interfacial width was calculated to be ~ 0.4nm. Carbonyl groups at the interfacial region of the polymer slab were found to prefer to form smaller angles with the interface normal to assist the formation of hydrogen bonds. The interfacial tension of the latex/water interface was found to decrease with the increase of the temperature. To validate the accuracy of the model, Interfacial tension of poly butyl acrylate was also calculated and found to agree well with the experimental data (within 10%) at various temperatures. OPLS and CHARMM force fields were used for all the simulations, and their predications of the interfacial properties of the system are similar. A coarse grained model based on MARTINI of the latex/water interface was also built and validated against the atomistic simulations.

(179d) Hydrogen Storage in Hydrates: Evaluation of Different Force-Fields Used in Monte Carlo Simulations

Papadimitriou, N. I.,
Tsimpanogiannis, I. N., National Center for Scientific Research
Economou, I. G., National Center for Scientific Research "Demokritos"
Stubos, A. K., National Center for Scientific Research "Demokritos"
Originally presented on: 11/4/2013 16:30:00 - 16:55:00

Clathrate hydrates have been examined as storage materials for hydrogen or other “energy-carrier” gases such as methane, as a result of having a number of advantages compared to other materials, including among others: operating at moderate temperatures, low cost, complete reversibility of the formation/dissociation process, minimal environmental hazards, having mainly water as a by-product. Initially the pure and binary hydrogen hydrates with structure II were experimentally confirmed [1], followed by the binary hydrate with structure H [2], and finally the structure I was obtained experimentally [3].

A requirement that needs to be met for a solid material to be suitable for practical applications in gas storage is high gravimetric and volumetric gas content. Therefore, in addition to performing experimental measurements for confirmation, it is essential to develop accurate theoretical/ numerical methods for the storage-capacity calculation of hydrates.

By assuming a fixed geometry for the hydrate lattice it allows the hydrate formation to be described as a process of gas storage in a solid material. Grand Canonical Monte Carlo (GCMC) methods have been used extensively to simulate gas adsorption in solid materials [4]. Therefore, the hydrate cavity occupancies, could be calculated from the GCMC simulations at various temperatures and pressures, and the results can be fitted to Langmuir-type functions [5].

In this study, an extensive series of GCMC simulations for the case of all the known hydrate structures, sI, sII, and sH that hydrogen is known to form, will be performed. During the simulations a number of water force-fields will be examined regarding their effect on the storage capacity of hydrates. In particular, the following popular water force-fields will be considered: SPC/E, TIP4P, TIP4P/Ice, and TIP5P. The Langmuir constants for each type of cavity and hydrate structure will be reported as a function of temperature and pressure (i.e., fugacities). Therefore, the storage-capacity of the different hydrate structures can be calculated.

References

[1] Mao, W. L.; Mao, H. K.; Goncharov, A. F.; Struzhkin, V. V. Guo, Q.; Hu, J.; Shu, J.; Hemley, R. J.; Somayazulu, M.; Zhao, Y. *Science* **2002**, *297*, 2247.

[2] Duarte, A. R. C.; Shariati, A.; Rovetto, L. J.; Peters, C. J. *J. Phys. Chem. B* **2008**, *112*, 1888.

[3] Grim, R. G.; Kerkar, P. B.; Shebowich, M.; Arias, M.; Sloan, E. D.; Koh, C. A.; Sum, A. K. *J. Phys. Chem. B* **2012**, *116*, 18557.

[4] Allen, M. P.; Tildesley, D. J. Computer Simulation of Liquids, Oxford University Press: New York, 1987.

[5] Papadimitriou, N. I.; Tsimpanogiannis, I. N.; Papaioannou, A. Th.; Stubos, A. K. *J. Phys. Chem. B* **2008**, *112*, 10294.

(179e) Atomistic Simulation Studies of the Enthalpy of Mixing of Systems Containing Aromatic and Aliphatic Hydrocarbon Oligomers

Rigby, D., Materials Design, Inc
Taylor, D. J. R.,
Originally presented on: 11/4/2013 16:55:00 - 17:20:00

Using a direct molecular dynamics simulation approach to obtain accurate measurements of the enthalpy of mixing in polymer blend systems is appealing because of its simplicity, although in practice a number of obstacles must be overcome. Specifically, whilst the density of simple small molecule liquid systems is often known accurately, this is seldom the case for systems containing oligomer or high polymer, and must be determined precisely and accurately from the simulation. This in turn implies that a high accuracy force field should be available. Moreover, since a variety of chemically distinct polymers will often be of interest, broad coverage of chemical functional groups must also be provided. A second series of potential difficulties involves the efficiency of sampling phase space during the simulation. Again, for small molecule systems this is generally not a problem, though as chain length increases sampling all representative regions of phase space within a single simulation becomes problematic, in particular because the broad range of energetically accessible conformations cannot be sampled. Consequently contributions to the mixing energy from changes in conformational energy as a function of composition, which may not always be negligible, become progressively more difficult to measure. Finally, many polymer blend systems of interest contain high polymer of molecular weights too large to be modeled directly. Fortunately, the mixing energy per unit volume tends towards a constant value as the molecular weight increases, and consequently the question that must be addressed in simulations is whether the range of accessible molecular weights is sufficient to reach the high molecular weight limit, or close enough to be obtained by extrapolation of oligomer data.

In view of advances in quality of available force fields and efficiency of readily available highly parallel simulation codes developed in recent years, it is of interest to reassess the accuracy and precision with which enthalpies of mixing in polymer-containing systems can routinely be predicted. In this work, we accordingly report on a study of aromatic-aliphatic mixture systems for which experimental data is widely available for both low molecular weight mixtures and higher molecular weight polymer. Specifically, we utilize a recently-refined version of the PCFF force field developed in this laboratory (denoted PCFF+) [1,2] and the Sandia LAMMPS simulation program [3], and consider systems which serve as models for the extensively-studied styrene-butadiene polymer mixture systems, beginning with a series of calculations of enthalpies of mixing in arene-alkane and arene-alkene systems, where results can be compared directly with experiment. Following this, we construct a series of short oligomer models that mimic the polystyrene-polybutadiene system and compute the (mostly unknown) densities and heats of mixing. Finally, we consider how the oligomer data can be extrapolated to higher molecular weights and compared with mixing enthalpies deduced from cloud point curves, subject to the assumption that the non-combinatorial contribution to the free energy of mixing can be neglected, as proposed experimentally for these systems.

References

1. see, for example, Sun, H., Mumby, S.J., Maple, J.R. and Hagler, A.T., J. Am. Chem. Soc. **116**, 2978 (1994)
2. Materials Design, Inc., Angel Fire, New Mexico, USA.
3. Plimpton, S., J. Comp. Phys. **117**, 1 (1995)

(179f) Vapor Liquid Equilibrium of Phenylacetylcarbinol

Madakashira, H., Indian Institute of Technology Bombay
Adhikari, J., Indian Institute of Technology, Bombay.
K, Y. R., Indian Institute of Chemical Technology
Originally presented on: 11/4/2013 17:20:00 - 17:45:00

R-Phenylacetylcarbinol (PAC), a chiral precursor, is an active pharmaceutical ingredient in manufacturing of various drugs such as ephedrine, adrenaline, etc. It is currently produced by means of biotransformation of benzaldehyde and pyruvate through whole cell fermentation using baker's yeast.[1] To achieve a better partitioning of PAC from aqueous to organic phase in the final step of the production process, it is important to know the physical and thermodynamic properties such as density, vapor pressure, boiling point, critical temperature, critical pressure, heat of vaporization, etc. As the only experimental data point available for PAC in literature[2] is the normal boiling point, we report the physical and thermodynamic properties as predicted through structure property correlations and molecular simulations. This is the first step towards predicting properties of solutions of PAC with other compounds present in the reactor.

Qualitative structure property correlations facilitate the prediction of physical properties from the structure of a molecule. A group contribution based Marrero-Gani (MG) method[3] has been employed to determine the boiling point and the critical properties of PAC. The normal boiling point obtained from MG method was around 75 K higher than the experimentally reported value of 479 K. Further, by making use of the predicted critical property data, the vapor liquid equilibrium has been determined using various equations of state[4] such as, Soave-Redlich-Kwong, Peng-Robinson, and volume translated Peng-Robinson.

Molecular simulations give a quantitative insight on various physical and thermodynamic properties within the limitations of the forcefield and its parameters. Constant volume Gibbs ensemble Monte Carlo (GEMC-NVT) simulations have been performed using MCCCS *towhee* package[5] in a temperature range of 350 K to 600 K, to determine pure component VLE and critical point of PAC. The transferable potentials for phase equilibria -united atom (TraPPE-UA) forcefield[6] developed by the Siepmann's group has been employed in the current study.The missing torsional profiles of various groups were obtained from density functional theory analysis at the B3PW91/6-31++G(d,p) level of theory using Gaussian03 package[7]. The boiling point predicted was determined to be 510 K, which lies between the prediction of the MG method and that reported from experiment. The VLE from GEMC-NVT is compared with the equation of state approach. The liquid and vapor densities at coexistence and the critical point data from both the methods are found to be in good agreement. The properties predicted include liquid and vapor coexistence densities, enthalpy of vaporization, saturation pressure, boiling point and critical properties.

References

1. Rosche, B., Sandford, V., Breuer, M., Hauer, B. & Rogers, P. L. Enhanced production of R-phenylacetylcarbinol (R-PAC) through enzymatic biotransformation. *Journal of Molecular Catalysis B: Enzymatic* **19-20**, 109–115 (2002).

2. Chapman & Hall. *Dictionary of organic compounds*. 6th edition, CRC press: New York, (1995).

3. Marrero, J. & Gani, R. Group-contribution based estimation of pure component properties. *Science* **184**, 183–208 (2001).

4. Schmid, B. & Gmehling, J. From van der Waals to VTPR: The systematic improvement of the van der Waals equation of state. *The Journal of Supercritical Fluids* **55**, 438–447 (2010).

5. MCCCS Towhee. Available at: http://towhee.sourceforge.net.

6. Keasler, S. J., Charan, S. M., Wick, C. D., Economou, I. G. & Siepmann, J. I. Transferable Potentials for Phase Equilibria − United Atom Description of Five- and Six-Membered Cyclic Alkanes and Ethers. *The Journal of Physical Chemistry B* ***116**, 11234-11246* (2012).

7. M. J. Frisch et al., GAUSSIAN 03, Revision C.01, Gaussian, Inc., Walling- ford, CT, 2004.

(214c) Molecular Interaction of Erythrosine B and Its Variations With Amyloid β Oligomer: Molecular Modeling Approach

Koo, D.,
Lee, J., Georgia Institute of Technology
Pyo, E.,
Jang, S. S., Georgia Institute of Technology
Kwon, I., University of Virginia

Molecular Interaction of Erythrosine B and its variations with Amyloid β Oligomer: Molecular Modeling Approach

Doyeon Koo,* Juho Lee,* Euisun Pyo and Seung Soon Jang

Computational NanoBioTechnology (CNBT) Laboratory
School of Materials Science and Engineering, Georgia Institute of Technology

771 Ferst Drive, Atlanta, GA 30332-0245

Inchan Kwon

Department of Chemical Engineering, University of Virginia

102 Engineers' Way, PO Box 400741
Charlottesville, VA 22904-4741

Alzheimer disease (AD) is the most common type of neurodegenerative disease which causes memory dysfunction, impaired judgment, and disorientation.[1] Although recent research has revealed a major pathological hallmark of AD, permanent treatment for AD is yet to be established for clinical use. Based on amyloid hypothesis, the AD treatment can be developed by investigating the formation of neuritic plaques composed of amyloid-β (Aβ) peptide fibrils from the cleavage of amyloid precursor protein (APP) by β- and γ-secretases[2]. Based on recent studies on AD, it has been discovered that small size of oligomer structure of Aβ42 peptide is known to be more toxic than fibril structure. Also, there is a difference in toxicity and aggregation between Aβ40 and Aβ42 peptides.[3]Therefore, the key point in finding an effective drug is the drug efficacy in dissociating the Aβ 42 peptides and preventing its aggregation.

In this molecular modeling study, Erythrosine B (ER), a Food and Drug Administration (FDA)-approved red food dye and its variations have been investigated to understand its mechanisms to hinder the aggregation of the Aβ peptides .[4] The most favorite binding sites and the relative binding affinity were evaluated by Autodock developed by the Scripps Institute, for monomeric strands and fiber units of Aβ40 and Aβ42, and subsequently the molecular dynamics simulations were performed to monitor the structural evolution of systems.

* They have equal contribution to this study.

References

[1] Alzheimer's Association. (2011) 2011 Alzheimer's Disease Facts and Figures. *Alzheimer's & Dementia.* 7(2): 21-23.

[2] Masters, C.L. et al.(1985) Amyloid plaque core protein in Alzheimer disease and Down syndrome. *Proc.Natl. Acad. Sci. USA* 82, 4245–4249

[3] Ahmed, M. Davis, J. Aucoin, D. et all. (2010) Structural conversion of neurotoxic amyloid- β1-42 oligomers to fibrils. *Nature structural & Molecular biology.* 17. 561-567

[4] Wong,H. Kwon, I. (2011) Xanthene Food Dye, as a Modulator of Alzheimer's Disease Amyloid-beta Peptide Aggregation and the Associated Impaired Neuronal Cell Function. *PLos One.* 6(10).

[5] Wong, H. Qi,W. Choi,H. Kwon, I. Fernandez, E.(2011) A Safe, Blood-Brain Barrier Permeable Triphenylmethane Dye Inhibits Amyloid-β Neurotoxicity by Generating Nontoxic Aggregates. *ACS Chem. Neurosci.* 2. 645–657

(214d) DNA At the Nanoscale: Interactions With Proteins, Polycations, and Surfaces

Elder, R. M., University of Colorado at Boulder
Jayaraman, A., University of Colorado, Boulder

Developing a fundamental understanding of the interactions of DNA with surfaces and with other biomolecules is important for numerous emerging biomedical and biomaterials technologies. In this poster we present three systems where we use molecular simulations to study the thermodynamics and molecular-level structural details of these interactions. In the first study, we examine how DNA is affected by two platinum-based anticancer drugs that, by design, covalently bind to and damage DNA to slow the growth of tumor cells. Recognition of this drug-DNA damage by repair proteins leads to drug-resistant tumors. We find that differences in the drug-DNA structure, and in the free energy of deforming the drug-DNA molecule during protein binding, explain differences in protein recognition of the anticancer drugs [1,2]. In the second study, we examine the behavior of single-stranded DNA (ssDNA) near model hydrophobic and hydrophilic surfaces, which is important for technologies based on DNA self-assembly. We find that ssDNA adsorbs strongly to both surfaces, although different types of interactions between the ssDNA, the surface, and water molecules govern adsorption [3]. In the third study, we examine the interactions between DNA and a novel class of polycations, which are effective biomaterials for gene delivery. We reveal molecular-level interactions in polycation-DNA binding using simulations with multiple levels of detail, thus complementing experimental data with otherwise inaccessible details such as binding free energy, local charge neutralization, and the structure of polycation-DNA complexes [4-6]. Most importantly, we have found that a close coupling of simulations and experiments enriches and accelerates biomedical and biomaterials research.

[1] Elder, R. M.; Jayaraman, A., Role of structure and dynamics of DNA with cisplatin and oxaliplatin adducts in various sequence contexts on binding of HMGB1a. *Mol. Sim.* 2012, 38, (10), 793-808.

[2] Elder, R. M.; Jayaraman, A., Sequence-Specific Recognition of Cancer Drug-DNA Adducts by HMGB1a Repair Protein. *Biophys. J.* 2012, 102, (10), 2331-2338.

[3] Elder, R. M.; Jayaraman, A., Structure and Thermodynamics of ssDNA Oligomers near Hydrophobic and Hydrophilic Surfaces: A Molecular Simulation Study. (*submitted*).

[4] Elder, R. M.; Emrick, T.; Jayaraman, A., Understanding the Effect of Polylysine Architecture on DNA Binding Using Molecular Dynamics Simulations. *Biomacromolecules* 2011, 12, (11), 3870-3879.

[5] Elder, R. M.; Jayaraman, A., Coarse-grained simulation studies of effects of polycation architecture on structure of the polycation and polycation-polyanion complexes. *Macromolecules* 2012, 45, (19), 8083-8096.

[6] Elder, R. M.; Jayaraman, A., DNA Binding Affinity of Nuclear Localization Sequence-based Polycations. (*in preparation*).

(214a) Replica Exchange Dynamic Molecular Simulations of Cyclic and Linear Amphiphilic Homopolymer

Liu, L., Tulane University

Replica Exchange Dynamic Molecular simulations of Cyclic and Linear Amphiphilic Homopolymer

Lixin Liu, Henry S. Ashbaugh*, Steve W. Rick*, Scott Grayson*
Department of Chemical and Biomolecular Engineering, Tulane University, New Orleans, LA 70118

Abstract

Polymer-based micelles had been raised particular interests in application as drug delivery material because their spontaneously forming core-shell structure with hydrophobic core encapsulating solvent-compatible particles as long as hydrophilic corona imparting water-solubility upon the aggregation. In our research work, Molecular Dynamic (MD) simulations based on a novel Amphiphilic Homopolymer c-PHS-Amph first reported by Scott Grayson's lab were performed to investigate polymer's unique topologies as well as its resulting prominent solvent-dependent character, which would progressively shed a in-depth interpretation to the molecular driving force leading amphiphilic behavior. Since Macromolecule requiring relative large scale computational work, we have cooperated with Steve Rick's group to employ Replica Exchange Dynamic Simulation (REDs) to improve sampling space and determine the global polymer conformation. Through applying REDs technique, exchange among replicas split in large potential barriers occurred efficiently and successfully deliver linear and cyclic polymer's characteristic structure corresponding with different solvents' environment. Proximal radial distribution of solvent around hydrophobic arms and hydrophilic arms was performed to validate the thermodynamic signatures as well as used to reproduce the density profile of interacting solvents.

(214f) Metallocene-Functionalized Boron-Doped Nanocarbons for Electrochemical Applications

Zhang, Z., University of Alabama
Liu, H., University of Alabama
Turner, C. H., University of Alabama

We have conducted a first-principles density-functional theory study on the electronic and electrochemical properties of cyclopentadienyl-transition metal (CpTM, with TM = Fe, Co, Ni) complexes adsorbed on pristine and boron-doped single-walled carbon nanotubes (B-CNTs) and graphenes, which may serve as new electrochemical catalysts or redox-active materials for electrochemical sensing. By using first-principles DFT calculations and ab initio molecular dynamics, we predict significant structural stablization by using B-CNT substrates, and we identify promising potential applications as electrochemical catalysts. Relevant to the well-studied electrochemical donor-acceptor molecule ferrocene, new CpFe/B-CNT or graphene-based electrochemical redox active materials may benefit from both the redox activity of the ferrocene and the superior stability and conductivity of B-doped nanocarbons. Therefore, the redox potentials of these complexes within aqueous solutions as well as in acetonitrile (MeCN) are calculated to evaluate their potential applications in electrochemical sensing. The calculated redox potentials of our catalysts are similar to those of ferrocene molecules, and this indicates the possibility of using these materials as ferrocene substitutes for electrochemical applications. Moreover, the effects of different B-CNT support chirality, doping patterns, and Cp sidechains on the redox potentials are also benchmarked. In addition to the redox calculations, the electron transfer details of the CpFe/B-graphene complex during the redox reactions are clarified by analyzing the natural bond order partial charges and by performing deformation charge density analyses.

(214ai) Hydrogen-Bonding-Suppressed Backbiting Reactions in Radical Polymerization of Acrylates: A Computational Study of Solvent Effects

Liu, S., University of Pennsylvania
Srinivasan, S., Arkema
Grady, M. C., DuPont Experimental Station
Soroush, M., Drexel University
Rappe, A. M., University of Pennsylvania

The intramolecular chain transfer (backbiting) reaction is the key reaction in acrylate polymerization responsible for converting a chain-end secondary propagating radical (SPR) to a less reactive mid-chain radical (MCR) [1]. The propagation and termination reactions of MCRs give rise to branching. At elevated temperature, MCRs can undergo β-scission reaction at considerable rate producing polymers with lower molecular weight [2-5]. Therefore, the occurrence of backbiting reaction directly influences the branching levels and molecular weight of final polymer product. Recent experimental studies demonstrate that the backbiting reaction is hindered significantly in the solvent that is capable of hydrogen-bonding to acrylate [6, 7]. However, the exact nature of the hydrogen-bonding interaction and its effect on the kinetics of backbiting reactions have not been fully understood.

This work presents a first-principles density functional theory study using B3LYP/6-31G*, X3LYP/6-31G*, M06-2X/6-31G* and B97-D3/6-31G* to explore the effects of hydrogen-bonding on backbiting reactions in high-temperature spontaneous polymerization of methyl acrylates (MA). Two types of solvent, dimethyl sulfoxide (DMSO) and methanol (CH_3OH) are studied. The effect of solvent is investigated with both conductor-like polarizable continuum model (PCM) [8] and SMD solvation model [9]. Our results suggest that the application of implicit solvation models has no significant effect of on the kinetics of backbiting reactions. Explicit solvent molecules are introduced to explore the hydrogen-bonding interactions between solvent and acrylates. Our computational results indicate that the presence of DMSO and CH_3OH increases the energy barrier of 1:5 backbiting reactions. This study is foreseen to provide insights into the use of hydrogen-bonding interactions to control the branching levels and molecular weight of poly(acrylates).

References

[1] Junkers, T.; Barner-Kowollik, C. *Journal of Polymer Science Part a-Polymer Chemistry* **2008**, *46*, 7585.

[2] Quan, C. L.; Soroush, M.; Grady, M. C.; Hansen, J. E.; Simonsick, W. J. *Macromolecules* **2005**, *38*,

7619.

[3] Chiefari, J.; Jeffery, J.; Mayadunne, R. T. A.; Moad, G.; Rizzardo, E.; Thang, S. H. *Macromolecules* **1999**, *32*, 7700.

[4] Junkers, T.; Bennet, F.; Koo, S. P. S.; Barner-Kowollik, C. *Journal of Polymer Science Part a-Polymer Chemistry* **2008**, *46*, 3433.

[5] Yamada, B.; Oku, F.; Harada, T. *Journal of Polymer Science Part a-Polymer Chemistry* **2003**, *41*, 645.

[6] Liang, K.; Hutchinson, R. A.; Barth, J.; Samrock, S.; Buback, M. Macromolecules **2011**, *44*, 5843.

[7] Srinivasan, S.; Kalfas, G.; Petkovska, V. I.; Bruni, C.; Grady, M. C.; Soroush, M. *J. Appl. Polym. Sci.* **2010**, *118*, 1898.

[8] Miertuě, S.; Scrocco, E.; Tomasi, J. *Chem. Phys.* **1981**, 55, 117

[9] Marenich, A. V.; Christopher, M.; Cramer, J.; Truhlar, D. G. *J. Phys. Chem. B* **2009**, *113* 6378.

(214e) A Molecular Simulation Study of Dispersants and Oil Hydrocarbons At Atmospheric Air/Salt Water Interfaces

Zhang, Z.,
Liyana-Arachchi, T. P., Louisiana State University
Avij, P.,
Ehrenhauser, F. S., Louisiana State University
Valsaraj, K. T., Louisiana State University
Hung, F. R., Louisiana State University

The explosion of the Deepwater Horizon (DWH) drilling rig in the Gulf of Mexico in April 2010 led to the second largest oil spill in U.S. territory and caused substantial environmental damage. Around 4.4 million barrels of crude oil were released during the accident, and a large amount of Corexit oil dispersants were used (1.8 million gallons); therefore, a significant fraction of the oil released and the dispersant used during the DWH accident was able to reach the surface of the sea. According to recent studies, the volatile and some of the intermediate volatile organic compounds (VOCs and IVOCs) that reached the sea surface evaporated into the atmosphere and formed aerosols. However, the fate of other oil organics, such as the heavier IVOCs (compounds with 17-18 carbon atoms) and the semi-volatiles (SVOCs, compounds with 19-31 carbon atoms) that reached the sea surface have not been addressed so far. The hypothesis behind our studies is that these heavier IVOCs, the SVOCs and the dispersants that reached the sea surface would be ejected into the atmosphere by phenomena such as white caps (breaking waves) and bubble bursting on the sea surface. Here, we investigated the properties of n-alkanes (C15, C20 and C30) and three model dispersant compounds (representative of species present in the Corexit dispersants used in the oil spill), at the air/salt water interface by using potential of mean forces (PMF) calculations and classic molecular dynamic (MD) simulations. Our simulation results show that the PMF of n-alkanes and model dispersants show deep minima at the air/salt water interface; these minima become deeper as the chain length of the alkanes increase, and as the concentration of dispersant in the system increases. The fact that the n-alkanes and the dispersant have a strong thermodynamic preference to remain at the air/salt water interface suggest that they are more likely to be ejected to the atmosphere by sea surface processes such as breaking waves and bubble bursting. These simulations agree with our experimental measurements of the ejection rates of oil hydrocarbons and Corexit dispersants. These trends imply that aerosolization through bubble-bursting and breaking waves at the sea surface is a likely transport mechanism for the ejection of spilled oil hydrocarbons into the atmosphere.

(214h) From Electrons to Engines: Kinetic Modeling of Low-Temperature Hydrocarbon Oxidation and Applications in Engine and Atmospheric Chemistry

Jalan, A., Massachusetts Institute of Technology
Green, W. H. Jr., Massachusetts Institute of Technology

From Electrons to Engines:
Kinetic Modeling of Low-Temperature Hydrocarbon Oxidation and Applications in Engine and Atmospheric Chemistry

Amrit Jalan and William H. Green

Department of Chemical Engineering, 77 Massachusetts Avenue, Room E18-566A, Cambridge, MA 02139, USA E-mail: whgreen@mit.edu

The chemistry of hydrocarbon oxidation at temperatures below ~800 K lies at the heart of several technological and natural phenomena. Examples include:

1. **Aging/thermal stability:** Oxidative degradation of polymers, fuels, biological lipids, etc. which are directly responsible for clogging arteries and fuel injectors alike.
2. **Energy:** Clean engine concepts like homogeneous charge compression ignition (HCCI) which rely on low-temperature auto-ignition to lower soot formation and NO_x emissions.
3. **Climate:** Atmospheric fate of organic material including processes like alkene ozonolysis and secondary organic aerosol formation.
4. **Bulk Chemicals:** Industrial oxidation processes used to manufacture chemicals like nylon, terephthalic acid and phenol, typically performed at low temperatures in solvent media.

Consequently, low-temperature oxidation has been the subject of experimental investigations for many decades. However, it is difficult to understand these processes at a mechanistic level from experiments alone for reasons like complex product mixtures which are difficult to analyze (e.g. liquid phase oxidation) or inherent instability of key reaction intermediates (like the Criegee Intermediate (CI) in alkene ozonolysis). Over the past 20 years, developments in scientific computing have opened up computational chemistry and chemical engineering tools as an attractive option for exploring and elucidating the kinetics of these complex processes.

In this work, we combine high-level electronic structure theory, reaction rate theory, automated kinetic modeling and empirical correlations for multi-scale modeling of low-temperature liquid phase oxidation and its application in modeling deposit formation in diesel fuel-injectors. We demonstrate the application of molecular modeling and engineering in the following areas:

1. **Elementary reaction kinetics:**
   - Use of quantum chemistry to explore new pathways for carboxylic acid formation in liquid-phase oxidation confirming a 30-year old experimental hypothesis.
   - Applications of similar pathways in atmospheric reactions of the Criegee Intermediate with carbonyl compounds.
2. **Reaction network analysis:**
   Use of automatic network generation to identify and resolve discrepancies between textbook models and experimental data on liquid phase oxidation.
3. **Coupled kinetics/phase equilibrium**
   Combine elementary kinetics with empirical phase-separation and transport models to simulate and investigate the mechanism of deposit formation in fuel injectors.

(214i) Structure, Dynamics and Crystallization of Ionic Liquids in the Bulk and Under Confinement

He, X.,
Santiso, E. E., Massachusetts Institute of Technology
Hung, F. R., Louisiana State University

In this poster we show results from two directly related projects. In our first project, molecular dynamics simulations were used to study the structure and dynamics of the ionic liquid (IL) $[bmim^+][NTf_2^-]$ confined inside the ordered mesoporous carbons CMK-3 and CMK-5, which consist of nanorods and nanopipes made of amorphous carbon. This material exhibits interconnected nanopores, which are promising candidates for applications in electrochemical double-layer capacitor (EDLCs). Our results indicate that variables such as pore size, pore morphology, temperature and surface density of electrical charges have a profound influence on the structural and dynamical properties of the confined IL. Ions inside the pore form different layers, with the number of layers and liquid structure varying with pore size, pore morphology and surface charge density. The dynamics of the ions far away from the surface of CMK-3 and CMK-5 are significantly faster than that of the ions near the carbon surfaces; in particular, increases in surface charge density leads to a significant reduction in the mobility of the counterions in the layers near the carbon walls.

In our second project we focus on novel nanomaterials based on frozen ILs. These nanomaterials hold enormous promise, as they exhibit the highly tunable properties of ILs. However, the design of crystalline materials is complicated because many molecules can crystallize into different polymorphs, which frequently exhibit different physical properties that affect their suitability for specific applications. To gain a molecular-level understanding of how cations and anions in an IL organize themselves into crystal structures, here we performed computer simulations of crystallization of the IL $[dmim^+][Cl^-]$ in the bulk. We used the string method in collective variables to determine minimum free energy paths (MFEPs) for crystal nucleation. Novel order parameters based on the generalized pair distribution function were used to characterize the different states connecting the liquid and crystal phases along the MFEP. The calculations also provide insights about activation barriers and the mechanisms of nucleation in these systems.

(214j) Oil Dispersion Using Hydrophobins

Chen, Y.,
Russo, P.,
Hung, F. R., Louisiana State University

Molecular simulation studies were performed to explore the feasibility of using hydrophobins, a class of proteins, in the dispersion of oil. Hydrophobins are a class of proteins produced by filamentous fungi in soil. Preliminary experimental results from the Russo group with the hydrophobin Cerato ulmin (CU) suggests that these proteins can efficiently encapsulate oil and air in cylindrical structures, which implies a striking surface activity. These properties, when combined with the abundancy of these proteins, have motivated applications such as removal of diesel fuel from contaminated fresh water. Because of their ease of (biosynthetic) manufacture, it is possible to imagine stockpiling hydrophobins for potential use as 'natural' oil spill dispersants. However, a fundamental understanding of the interfacial properties of hydrophobins when they encapsulate oil (to form 'blobs') or gases (to form bubbles), is crucial to determine whether CU can be an efficient oil/gas dispersant. Here we investigated the properties of hydrophobins on air/water and oil/water interfaces using potential of mean force (PMF) calculations and classical molecular dynamics (MD) simulations, using both all-atom and coarse-grained representations. Relevant interfacial properties, such as free energies, density profiles, structure and stability of small blobs and bubbles are investigated and discussed in this study.

(214g) Confinement and Adsorption of Fluids in Nanoporous Materials From Molecular Dynamics Simulations: Influence On Membrane Separation Performance

Frentrup, H., Imperial College London
Muller, E. A., Imperial College London

Separation performance is expressed in terms of either mass throughput or selectivity, and it is the trade-off between these two paradigms that poses the primary challenge for membrane development. While selectivity is crucially determined by the energetic and geometric characteristics of the porous materials, mass throughput can be limited by the mass transfer rate through the selective barrier, the dynamics of the adsorbed layer, entrance effects and mainly by the transport dynamics of the confined fluid inside the pore [1].

The focus of this research lies on the influence of the latter aspects on membrane separations, namely on the effect that nanoscale confinement has on fluid transport and on separation behaviour of ultrathin-film membranes, where adsorption dynamics and entrance effects are prominent [2]. Under nanoscale confinement, fluids exhibit a highly complex behaviour, markedly distinct from their bulk conterparts. The distinction between diffusive and convective mass transport as well as a competition between the phenomenological description and a fully molecular understanding of mass transport [3] give rise to several unanswered questions with respect to fluid transport in porous media. Transport processes in porous materials are influenced by structural and thermodynamic effects and the effects are often superimposed which makes a clear distinction very difficult [4, 5].

It is common to characterise transport properties of confined fluids via the self-diffusion coefficient based on the Einstein diffusion model. The validity of this approach in the context of nano-confined fluids is very limited and it fails to capture important factors [6]. We showcase these limits for a popular reference case, a Lennard-Jones fluid confined in structured slit pores. The influence of the repulsive and attractive fluid-pore interactions, pore loading as well as steric confinement effects on transport is presented. Self-diffusion and collective diffusion coefficients from Equilibrium Molecular Dynamics (EMD), based on the Green-Kubo relations [7], and the thermodynamically related Fickian diffusivities, were calculated.

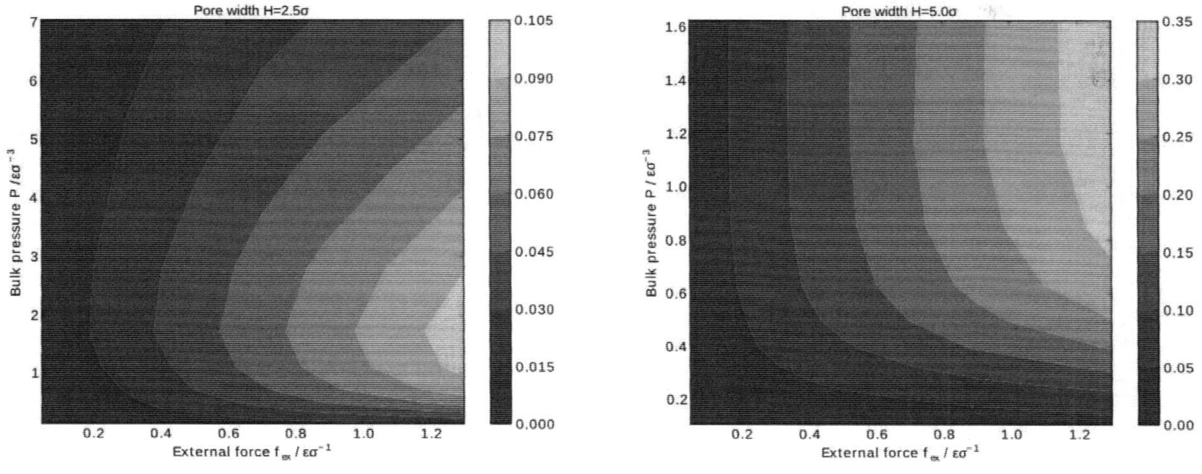

Figure 1: For the NEMD simulations, a LJ fluid in contact with a slit pore was perturbed by an external force at the boundary of the simulation box, inducing a density gradient and fluid flow. The steady state flux (in reduced units) as a function of bulk fluid pressure and magnitude of the external field is plotted for a narrow slit pore on the left (H=2.5) and a wide pore on the right (H=5.0) in the depiction above. Simulations were performed at a temperature of T=1.5.

We propose a boundary-driven Non-Equilibrium Molecular Dynamics (NEMD) algorithm in an attempt to calculate corresponding effective diffusion coefficients [8]. The non-equilibrium nature of the methodology allows us to study diffusion under steady state conditions and to be able to discriminate between the transport coefficients inherent of the porous matrices and the entrance and exit effects.

In an attempt to paint a holistic picture of single-component diffusion under confinement, we draw a direct comparison between transport diffusion coefficients calculated from EMD and effective diffusivities from NEMD for a range of pore loading and sizes, as well as varying fluid-pore interactions (c.f. Fig. 1). These results are put in contrast to theoretical models of mass transport in porous materials such as the Knudsen diffusion approach and predictions from hydrodynamics on the molecular level [9], where limitations of these theories become apparent. It stems from our results that a description of transport through porous membranes without due consideration of the either enhanced or restricted transport at the solid- bulk fluid interfaces leads to significant deviations from the expected overall transport.

With considerable effort going into fabricating membranes of an atomistic thickness [10], the limits on maximising mass throughput and selectivity simultaneously are furthermore explored. The selective behaviour of these ultra-thin membrane layers is seen to be dominated by the dynamics of the adsorbed layers which can , however, have significantly negative influence on the separation performance.

References

[1] W. M. Deen, *AIChE Journal*, **33**, 1409--1425 (1987).
[2] C. Huang, L. W. Kostiuk et al., *Phys. Chem. Chem. Phys.*, **10**, 186--92 (2008).
[3] D. Keffer, C. Gao, and B. Edwards, *J. Phys. Chem. B*, **109**, 5279 (2005).
[4] Q. Cai, M. Biggs, and N. Seaton, *Phys. Chem. Chem. Phys.*, **10**, 2519 (2008).
[5] R. Krishna, *J. Phys. Chem. C*, **113**, 19756 (2009).
[6] O. Jepps, S. Bhatia, and D. Searles, *Phys. Rev. Lett.*, **91**, 126102 (2003).
[7] G. Arya, H. Chang, and E. Maginn, *J. Chem. Phys.*, **115**, 8112 (2001).
[8] H. Frentrup, C. Avendaño, M. Horsch, A. Salih, and E. A. Müller, *Mol. Sim.*, **38**, 540 (2012).
[9] K. Travis and K. Gubbins, *J. Chem. Phys.*, **112**, 1984 (2000).
[10] Y. Han, Z. Xu, and C. Gao, *Adv. Funct. Mat.*, DOI:10.1002/adfm.201202601 (2013).

(214k) Enhancing the Surface Reactivity of SrxY1-XTiO3 Mixed Ionic and Electronic Conductor Toward Fuel Oxidation in SOFC: First-Principles Study

Ham, H. C., Korea Institute of Science and Technology
Kim, H. S., KIST
Yoon, S. P., KIST
Han, J., Korea Institute of Science and Technology
Nam, S. W., Korea Institute of Science and Technology

In recent years, the yttrium-doped strontium titanium oxide ($Sr_{1-x}Y_xTiO_3$) has received much attention as an alternative anode catalyst for solid oxide fuel cell (SOFC) due to its high sulfur/carbon resistance, good electronic/ionic conductivity and phase stability in typical SOFC anode operating conditions. However, $Sr_{1-x}Y_xTiO_3$has exhibited the poor electro-catalytic activity toward fuel (i.e. H_2 or hydrocarbon) oxidation. Thus, it is very imperative to understand the factors controlling the reactivity of $Sr_{1-x}Y_xTiO_3$ toward SOFC anode reactions for developing effective $Sr_{1-x}Y_xTiO_3$-based anode materials.

In this work, we will present some recent theoretical results we have on the role of surface oxygen vacancy formation energy in determining the kinetics of H_2 oxidation. Using spin-polarized density functional theory calculations for geometric and electronic structures, we find that the H_2 oxidation reaction [H_2+O(surface) $\rightarrow$ OH+OH $\rightarrow$ H_2O+ O(vacancy)] on the surface of $Sr_{1-x}Y_xTiO_3$ -based anode materials strongly depends on the oxygen vacancy formation energy. Our calculations also show the oxygen vacancy formation energy can be controlled by doping the transition-metal into the Ti lattice site of $Sr_{1-x}Y_xTiO_3$.

This work hints the importance of knowing how to properly tailor the surface reactivity of oxide-based anode materials for achieving wanted electro-catalytic activity for SOFC application.

(214l) Using Monte Carlo Simulations to Study Bulk and Interfacial Properties of Ionic Liquids

Rane, K. S., University at Buffalo, The State University of New York
Errington, J. R., University at Buffalo, The State University of New York

Ionic liquids show interesting thermo-physical properties due to their unique chemical nature which is characterized by comparative strengths of coulombic and dispersion interactions. In this poster we discuss how molecular simulations are used to study wetting properties of these fluids on solid substrates over a range of conditions. We study wetting phenomena at liquid-vapor saturation conditions and therefore understanding the phase coexistence properties of these fluids is also important. Wetting properties are computed using a free energy based approach that involves performing Monte Carlo simulations to grow a liquid or vapor film on the solid surface of interest. The properties computed from these simulations include interfacial tensions, surface excess energies, and entropies. Expanded ensemble simulations are used to scan a broad range of temperatures and substrate strengths, which enables us to study systems in different wetting regimes. Results are presented for the wetting behavior of different models of ionic liquids on solid substrates. To calculate liquid-vapor saturation properties of ionic liquids over a broad temperature range we use a combination of direct grand canonical and isothermal-isobaric temperature expanded ensemble Monte Carlo simulations. We also show how advanced simulation tools are used to address the challenges generally encountered in simulating complex molecules like those of ionic liquids. We present results for homologous series 1-alkyl-3-methylimidazolium bis(trifluoromethylsulfonyl)imide ([$C_n$mim][NTf_2]) from room temperatures to temperatures close to the estimated critical points.

(214m) Coarse-Grained Modeling of the Phase Behavior of Thermodynamically Small Particle Assemblies

Sehgal, R. M., University of Massachusetts Amherst
Ford, D., University of Massachusetts Amherst
Maroudas, D., University of Massachusetts Amherst

The phase behavior of systems which are termed thermodynamically small has been the subject of intensive theoretical study over the past two decades. These finite systems consist of a very small number of interacting particles, ~10 to ~100, and, as such, they are far removed from the infinite limit in the sense of traditional macroscopic thermodynamics. In such small systems, the nature of thermodynamically stable phases and the corresponding phase transitions are of particular interest. Developing a fundamental understanding of the phase behavior of these systems has direct application for the self- and directed-assembly of colloidal particles into structures within materials and devices as an emerging paradigm with wide-ranging technological impact. In addition, these finite-size particle clusters may serve as model systems for the types of phase behavior expected and observed in other complex systems, such as proteins, glasses, and gels.

In this poster presentation, we report results of a systematic investigation of phase behavior in two thermodynamically small systems, namely, finite assemblies of colloidal particles interacting via a hard-core and a depletion-attraction potential and the 38-member Lennard-Jones (LJ_{38}) particle cluster. In the colloidal system, we have studied the order-to-disorder phase behavior over a broad range of parameter space with parameters that include the system size, expressed by the number of particles N in the colloidal assembly, and the inter-particle interaction strength, which is controlled by the depletant osmotic pressure Π/kT. In the LJ_{38} cluster, we have focused on a single parameter, the system temperature kT/ε, and studied its effects on polymorphic solid-solid and melting solid-fluid phase transitions.

To provide a description of the phase behavior of these systems, we have constructed free-energy landscapes (FELs) based on Monte Carlo umbrella sampling (MC-US). A coarse-grained representation of the thermodynamically small system is required for the implementation of MC-US and the construction of meaningful FELs. This coarse-grained model allows for reducing the description of the system from a $3N$-dimensional ($3N$-D) representation into a much coarser description with a much lower dimensionality. In order to perform this dimensionality reduction rigorously and systematically, we have applied the diffusion mapping approach to both systems of interest, with data sets provided by Brownian- or molecular-dynamics simulations for the colloidal and LJ_{38} clusters, respectively. The effects of varying the system parameters on the FELs have been examined systematically.

In the colloidal particle assemblies, we find that only a single fluid-like phase is stable for very small clusters. However, as the cluster size increases, a second ordered phase emerges in coexistence with the fluid-like phase; this second stable phase at increased cluster size is a crystalline phase. The onset of stability of this crystalline phase corresponds to the size of a critical crystalline nucleus and marks the onset of crystallization in colloidal particle assemblies. In the LJ_{38} cluster, we find that, at very low temperatures, the system only samples its minimum-energy, octahedral configuration. However, as the temperature increases, the system undergoes a polymorphic transition between two different solid phases followed by an order-to-disorder, solid-to-fluid melting-like transition at even higher temperatures.

(214n) Effect of Mechanical Deformation On Structure in Nafion 117 From Molecular Dynamics Simulations

Abu-Hakmeh, K., Georgia Institute of Technology
Piper, N.,
Cheevaphantusri, N.,
Chun, B., Georgia Institute of Technology
Jang, S. S., Georgia Institute of Technology

Nafion is commonly used in polymer electrolyte membrane fuel cells (PEMFC) due to its high proton conductivity and thermal, chemical, and mechanical stability. So far, studies of Nafion have heavily focused on the relationship between nanophase-segregated structure and transport, revealing that proton transport is enhanced in more developed nanophase-segregated structures. Recently, hydrated Nafion membrane was uniaxially extended in experiment, which showed an improvement of transport properties. In this study, mechanical properties of Nafion membrane are investigated using MD simulation methods. Specifically, the deformation of Nafion is implemented with various water contents to achieve its effect on the structure and transport of hydrated Nafion membrane. First, Nafion 117 membrane is equilibrated with certain levels of hydration through an annealing procedure. The simulated characteristic correlation length and the diffusion coefficient of water and hydronium are analyzed for comparison with those observed in experiments. Then, the equilibrated Nafion membrane is deformed uniaxially with a constant strain rate up to 300% of strain. The change in nanophase-segregation of hydrated Nafion during the deformation process is characterized using characteristic correlation length as well as the pair correlation function in order to achieve fundamental understanding of the relationship of such structural change as a function of strain with the proton transport.

(214o) Atomic-Scale Modelling of Hydrogen Storage in a Series of Metal-Organic Frameworks

Andersson, S., SINTEF Materials and Chemistry
Larsson, P. E.,

Atomic-scale modelling of hydrogen storage in a series of metal-organic frameworks

Stefan Andersson and Per-Erik Larsson

SINTEF Materials and Chemistry, P.O. Box 4760, 7465 Trondheim, Norway

Atomic-scale modelling is a powerful tool able to predict adsorption and transport properties of gases in metal-organic frameworks (MOFs). In this work we predict the key adsorption and transport properties for hydrogen storage applications, by investigating a series of recently synthesized zirconium based MOFs. These MOFs consists of an inorganic brick ($Zr_6O_4(OH)_4(CO_2)_{12}$) that binds to organic linkers (1,4-benzene-dicarboxylate, 4,4'-biphenyl-dicarboxylate and terphenyl dicarboxylate) to form the three MOFs: UiO-66, UiO-67 and UiO-68. These materials are of particular interest because of their, relative to other MOFs, superior thermal stability (being stable to above 500 °C). For the UiO series of MOFs we will present results from Grand Canonical Monte Carlo and Molecular Dynamics simulations of key parameters such as adsorption isotherms and diffusion coefficients. Excellent agreement is found with available experiments in several cases. In addition, the atomic-scale modelling provides molecular level details that are difficult to study with experiments.

(214p) Simulated Thermal Annealing of Low-k Organosilicate Glass Materials Using the Reax-FF Force Field

Raymunt, A.,

The continued shrinking of microelectronic devices has led to the need for lower dielectric constant (low-k) materials in back-end interconnect structures to reduce resistive-capacitive delay in signal propagation. In order to reach lower dielectric constants traditional materials are used as a starting point, but are fabricated to promote the formation of pores. Higher levels of porosity reduce the dielectric constant, but also result in a loss of mechanical strength.

We have modeled low-k organosilicate glasses using molecular dynamics techniques in order to understand the relationship between the material's structure, porosity, dielectric constant, and mechanical strength. In this paper, we use the REAX-FF force field to model the material's atomic interactions. This force field is unusual because it allows for the breaking and forming of bonds during the simulation. This capability allows us to follow the evolution of the bonding environment of the material's structure as we change the thermal environment of the system. Ultra-rapid thermal annealing is of interest since it offers the potential to strengthen the material with minimal increase in dielectric constant.

We begin by establishing a starting REAX-FF model of these SiCOH-type materials with the appropriate composition and bulk properties to match an experimental film. We then simulate thermal annealing of this material, after establishing a simulation-derived glass transition temperature (T_g) (~1000 K) to use as a point of reference. By comparing the structure at T_g with the starting structure, as well as with the structures at various temperatures along the way, we are able to link temperature to structural and property changes. We will show how the structure evolves with increased temperature to eventually result in a point of pore collapse and higher dielectric constants. We also suggest processing paths to lead to a balance between mechanical strength and dielectric constant and draw comparisons to experimental results.

(214r) Molecular Dynamics Simulation of Dual Polyelectrolyte Coagulation With Natural Organic Matters in Aqueous System
Choi, Y. J.,

Molecular Dynamic Simulation of Dual Polyelectrolyte Coagulation with Natural Organic Matters in Aqueous System

Yoon Jung Choi,[a] Hyemin Lee,[a] Jiwon Jung,[a] Sang Hyuck Park[b] and Seung Soon Jang[a]

[a] School of Materials Science and Engineering, Georgia Institute of Technology, 771 Ferst Drive, Atlanta, GA 30332-0245

[b] School of Science & Technology, Georgia Gwinnett College, 1000 University Center Lane, Lawrenceville, GA 30043

Abstract

In this research, water treatment using dual-coagulants is examined. Through Molecular Dynamics (MD) simulation, the coagulation process with natural organic matter (NOM) is studied and analyzed at molecular level. Two of polyelectrolytes, Polydiallyldimethylammonium chloride (PDADMAC) and Polyacrylic Acid (PA), are used as two coagulants. Those dual-coagulants are simulated with Humic Acid trimer with Temple-Northeastern-Birmingham (TNB) model. During the simulation, PDADMAC forms complexes with NOM in water phase, which are then further coagulated and flocculated with the secondary polyelectrolyte of the opposite charge. This simulation with two coagulants and NOM was investigated under various conditions of different solvent, ionic strength, and molecular weight of polyelectrolytes. Throughout the extensive simulation and analysis, it was found that the binding of PDADMAC with NOM is mainly due to electrostatic binding. It was also found that, even after the PDADMAC aggregates with NOMs, PDADMAC can still interact with PA through electrostatic interaction.

(214q) Structure and Transport Property of Sodium Dodecyl Surfate Micelle in Water Phase: Molecular Dynamics Simulation Approach

Chun, B., Georgia Institute of Technology
Jang, S. S., Georgia Institute of Technology

Structure and Transport Property of Sodium Dodecyl Surfate Micelle in Water Phase: Molecular Dynamics Simulation Approach

Byeong Jae Chun and Seung Soon Jang

Computational NanoBio Technology Laboratory, School of Materials Science and Engineering, Georgia Institute of Technology, 771 Ferst Drive NW, Atlanta, GA 30332-0245

We investigate the structure and transport property of micelle consists of sodium dodecyl surfate (SDS) in water phase using full atomistic molecular dynamics (MD) simulation. For this, first, SDS is geometry-optimized using B3LYP and 6-31G** and then Mulliken population analysis is performed for atomic charges. Â Second, a spherical self-assembled micelle is prepared by arranging SDS molecules to have a centrosymmetric inner structure and subsequently is located in water phase. By performing MD simulation, the structure and stability of micelle are characterized by 1) radius of gyration; 2) density distribution; 3) Connolly surface area; 4) distribution of sodium atoms; 5) formation energy. The simulated structure and energy will be compared to experimental ones in order to validate our simulation protocols as well as force field. Once the equilibrated SDS micelle is validated, the transport of various small atomic and molecular species through micelle is analyzed using potential of mean force (PMF) to characterize the permeability of SDS micelle in terms of free energy barrier. We believe this study provides a molecular-level understanding of micelle for its applications towards molecular carriers and molecular reactors.

(214s) SOx Binding On Fly Ash Components for Oxy-Coal Combustion

Galloway, B., University of South Carolina
Padak, B., University of South Carolina

The United States is responsible for emitting 1.5 billion tons of carbon dioxide (CO_2) per year from the combustion of coal in the generation of electricity. These CO_2 emissions have been garnering greater attention in recent years due to CO_2's classification as a greenhouse gas and its role in global warming. One technology being developed to combat this problem is oxy-coal combustion where oxygen, not air, is used as the oxidizer in the boiler. The elimination of nitrogen produces a concentrated CO_2stream that can be captured and compressed for transport and sequestration. For the existing conventional power plants to be retrofitted, the operational and environmental impacts of oxy-coal combustion need to be evaluated.

Previous research has shown that sulfur dioxide (SO_2) emissions can be lowered by employing oxy-coal technology. The higher oxygen concentration and recycle stream leads to greater oxidation of SO_2 to sulfur trioxide (SO_3) resulting in sulfur retention on fly ash and ash deposits in the furnace. While the increased sulfur retention on the ash particles decreases SO_2 emissions, it creates problems utilizing the fly ash for cement and concrete production. To address this problem and aid in making oxy-coal combustion a viable technology, one must first examine the binding of sulfur oxides (SO_X) on the various components of fly ash, e.g. calcium oxide (CaO), magnesium oxide (MgO) etc. To better understand the sulfur retention mechanism, density functional theory (DFT) calculations were performed using plane wave basis sets to investigate these earth metal oxides. A slab model was used to simulate clean CaO(100), MgO(100), Na_2O(100), and K_2O(100) surfaces for SO_3 binding. Energies and density-of-states were calculated for a variety of binding sites and adsorbate orientations to examine the binding mechanism and dominant species on these surfaces.

(214aj) Free Energy Based Method for Calculating Elastic Constants of Liquid Crystals

Joshi, A. A., University of Wisconsin
Whitmer, J. K., University of Wisconsin
Guzman, O., Universidad Autonoma Metropolitana-Iztapalapa
DePablo, J. J., University of Wisconsin-Madison

There is considerable interest in designing liquid crystals capable of yielding specific morphological responses in confined environments, including capillaries and droplets. The morphology of a liquid crystal is largely dictated by the elastic constants, which are difficult to measure and are only available for a handful of substances. In this work, a first-principles based method is proposed to calculate the Frank elastic constants of nematic liquid crystals directly from atomistic models. These include the standard splay, twist and bend deformations, and the often-ignored but important saddle-splay constant. The proposed method is validated using a well-studied Gay-Berne(3,5,2,1) model; we examine the effects of temperature and system size on the elastic constants in the nematic and smectic phases. We find that our measurements of splay, twist, and bend elastic constants are consistent with previous estimates for the nematic phase. We further outline the implementation of our approach for the saddle-splay elastic constant, and find it to have a value at the limits of the Ericksen inequalities. We then proceed to report results for the elastic constants of atomistic models of liquid crystals, including 5CB and 8CB, and show that the values predicted by our approach are consistent with a subset of the available but limited experimental literature.

(214t) Scft Simulation of Rod-Coil Block Copolymer Nanocomposites

Trask, L., Iowa State University
Cochran, E., Iowa State University

Computer simulations are performed for rod-coil block copolymer systems containing embedded nanoparticles. Due to the complexity of the rod-coil block copolymer phase space, systems of interest are restricted to the lamellar phases; these include the smectic and nematic phases. Self-consistent field theory is used to model the rod-coil block copolymer interactions and behavior. The simulations are performed for varying Maier-Saupe interaction strengths, rod lengths, different particle sizes, and particle affinities.

(214ak) Polyelectrolyte Interactions: Theory and Simulation

Sammalkorpi, M., Aalto University
Antila, H., Aalto University
Van Tassel, P. R., Yale University

Polyelectrolyte interactions govern many natural and technological processes. Recent experiments and simulations suggest like charged polymers to attract one another under certain conditions, owing to preferential counter-ion positioning between the two chains. Such attraction may have significant implications to nucleic acid delivery and polyelectrolyte film formation. We address here the problem of like-charged polymer attraction using a coarse grained model, consisting of two similarly charged aligned cylinders, oppositely charged spherical counter-ions, and spherical salt species. Employing Monte Carlo simulation, we show polymer-polymer attraction to correlate strongly with a Coulomb coupling parameter, defined as the ratio of the electrostatic interaction energy of two counter-ions at the cylinder surface to the thermal energy. We also find a significant influence of ion diameter in the regime where crowding induces ion positioning away from the cylinder surface. Two simple statistical mechanical models are proposed: a purely mean field model for larger counter-ion diameter, and a hybrid model for smaller counter-ion diameter where ions "condensed" to the cylinder surface are treated as a highly correlated 2-D one-component plasma, and ions away from the surface are treated in a mean field description. We discuss these results in light of recent experimental findings of significantly enhanced polyelectrolyte adsorption to electrified interfaces.

(214al) Free Energy of Solvated Salt Bridges: A Simulation and Experimental Study

White, A. D., University of Washington
Keefe, A. J., University of Washington
Nowinski, A. K., University of Washington
Shao, Q., University of Washington
Pfaendtner, J., University of Washington
Jiang, S., University of Washington

Charged amino acids are the most common on surfaces of proteins and understanding the interactions between these charged amino acids, salt bridging, is crucial for understanding protein-protein interactions. Previous simulations have been limited to implicit solvent or fixed binding geometry due to the sampling required for converged free energies. Using new accelerated sampling techniques, we have calculated salt bridge free energy surfaces in water and compared the results with NMR experiments. The simulations give binding free energies, quantitative ranking of salt bridging strength, and insights into the hydration of the salt bridges. The arginine-aspartate salt bridge was found to be the weakest and arginine-glutamate thestrongest, showing that arginine can discriminate between aspartate and glutamate, whereas the salt bridges with lysine are indistinguishable in their free energy. The salt bridging hydration is found to be complementary to salt bridge orientation with arginine having specific orientations.

(214u) Asphaltene Structural Models Validated By Density Functional Theory

Oliveira, J. S. C.,
Seidl, P. R., UFRJ
Stoyanov, S. R.,
Albuquerque, M. G., Federal University of Rio de Janeiro
Leal, K., Universidade Federal Fluminense

Asphaltenes – a major component of heavy petroleum – are a complexe mixture of mainly high molecular weight organic compounds, containing polyaromatic moieties tethered with aliphatic side chains. Asphaltenes are defined as the fraction of petroleum soluble in aromatic solvents (e.g., toluene) and insoluble in paraffinic ones (e.g., n-heptane).

Asphaltenes cause serious problems of precipitation during oil production and poisoning of catalyst during refining. These problems are particularly serious in countries like Brazil, where major oil reserves lie deep under the ocean, and Canada, where they are extracted from Oil Sands.

The chemical complexity of the asphaltene fraction of petroleum makes the elucidation of the structures of asphaltene molecules very difficult. Several analytical methods (e.g., NMR, XRD) have been employed to characterize asphaltene fractions from different petroleum samples.

Since the molecular structure of asphaltenes remains elusive, models that represent average asphaltene fractions, such as quantitative molecular representations [1] and synthetic model compounds [2,3] have been developed and employed to study the behavior and properties of asphaltenes.

In previous studies, we calculated the association energy of representative structures of asphaltenes, using molecular dynamics simulations in the presence of a solvent [4].

The objective of the current project is to validate a representative model of asphaltenes developed based on a correlation between experimental NMR and computational results on the association behavior of a series of reference model compounds [4].

The calculations have been performed using the ωB97X-D density functional with an augmented damped R^{-6} dispersion correction term and the basis set 6-31G(d,p), as implemented in the Gaussian 09 computational chemistry software suite [5].

The reference model compounds are hexabenzocoronene (HBC), hexa(n-hexyl)-HBC (C6-HBC), and pyrene (P). Experimental results [2,6] indicate that HBC or C6-HBC molecules are associated even in polar solvents, whereas pyrene do not associate. The energy of dimerization of the reference models will be compared with our previous models (A, B, C, and D) [4]. The dimerization energy is used as an association parameter.

The dimerization energy values calculated for HBC and pyrene show that these asphaltene models are associated in a vacuum. We intend to correlate the experimental and computational results and define a dimerization energy limit so as to predict the occurrence of dimerization.

The correlation shows that structures with a gain in dimerization energy over 20 kcal/mol are associated according to the experiment, whereas those with lower dimerization energy values are either weakly associated or do not associate.

References:

[1] Boek E.S. et al. (2009) Energy Fuels, 23 (3), 1209–1219.

[2] Tan X. et al. (2008) Energy & Fuels 22, 715-720.

[3] Costa L.M. et al. Phys. Chem. Chem. Phys. 2012, 14, 3922–3934;

[4] Oliveira, J.S.C. et al. (2010) Poster Presentation at the 11th International Conference on Petroleum Phase Behaviour and Fouling, PetroPhase, New York, NY, USA.

[5] Frisch M.J. et al. (2009) Gaussian 09, Revision A.1.

[6] Akbarzadeh K. et al. (2005) Energy & Fuels 19, 1268-1271.

(214v) GPU Accelerated Configurational Bias Monte Carlo Simulations of Branched Molecules

Mick, J. R., Wayne State University
Rushaidat, K. I., Wayne State University
Hailat, E., Wayne State University
Li, Y., Wayne State University
Schwiebert, L., Wayne State University
Potoff, J. J., Wayne State University

The simulation of systems containing over 100,000 atoms has become commonplace with the development open-source parallelized molecular dynamics simulation codes, such as LAMMPS and NAMD [1, 2]. In addition to outstanding scaling across CPUs, these codes, and others, such as HOOMD [3], have been written to utilize inexpensive graphics processors. For the simulation of processes at constant temperature and or pressure, molecular dynamics is clearly the method of choice, however, chemical or biological processes that occur at constant chemical potential require simulation in an ensemble that allows for fluctuations in the number of molecules. Simulation in open systems is performed typically with Monte Carlo simulations in either the grand canonical or Gibbs ensembles. However, the large system sizes [4, 5] required for biomolecular systems make such calculations computationally prohibitive using existing serial codes.

In this work, a multi-ensemble Monte Carlo simulation engine, GO-MC(GPU-Optimized Monte Carlo), capable of simulating systems of tens of thousands of molecules of arbitrary geometry on a desktop workstation is presented. The engine is built on past code to simulate linear molecules on the GPU [6, 7] and is written in C++ using NVIDIA Corp.'s Compute Unified Device Architecture (CUDA) API. Performance comparisons are made between GPU implementations of the coupled-decoupled and a coupled configurational bias methods [8-10]; as measured by acceptance rates and computation time on the device. The coupled approach, while inefficient on the CPU, shows promise on the GPU due to better thread balancing. Performance improvements due to the use of a nearest neighbor cell list [11] are discussed. Further code optimizations available on the NVIDIA Kepler architecture are described.

1. Brown, W.M., et al., *Implementing molecular dynamics on hybrid high performance computers - short range forces.* Computer Physics Communications, 2011. **182**(4): p. 898-911.

2. Phillips, J.C., J.E. Stone, and K. Schulten, *Adapting a message-driven parallel application to GPU-accelerated clusters*, in *Proceedings of the 2008 ACM/IEEE conference on Supercomputing*2008, IEEE Press: Austin, Texas. p. 1-9.

3. Nguyen, T.D., et al., *Rigid body constraints realized in massively-parallel molecular dynamics on*

graphics processing units. Computer Physics Communications, 2011. **182**(11): p. 2307-2313.

4. Freddolino, P.L., et al., *Molecular Dynamics Simulations of the Complete Satellite Tobacco Mosaic Virus.* Structure (London, England : 1993), 2006. **14**(3): p. 437-449.

5. Sanbonmatsu, K.Y., S. Joseph, and C.-S. Tung, *Simulating movement of tRNA into the ribosome during decoding.* Proceedings of the National Academy of Sciences of the United States of America, 2005. **102**(44): p. 15854-15859.

6. Jason R. Mick, K.R., Eyad Hailat, Vincent Russo, Loren Schwiebert, Jeffrey J. Potoff, *GPU Accelerated Configurational Bias Monte Carlo Simulations of Linear Alkanes.* Proceedings of the 2013 AIChE Annual Meeting, 2012: p. 51e.

7. Jason R. Mick, K.R., Eyad Hailat, Vincent Russo, Loren Schwiebert, Jeffrey J. Potoff, *GPU-accelerated Gibbs ensemble Monte Carlo simulations of Lennard-Jonesium"* Comp. Phys. Comm. (submitted, in review), 2013.

8. Frenkel, D. and B. Smit, *Understanding molecular simulation: from algorithms to applications*2002: Academic Press.

9. Martin, M.G. and J.I. Siepmann, *Novel Configurational-Bias Monte Carlo Method for Branched Molecules. Transferable Potentials for Phase Equilibria. 2. United-Atom Description of Branched Alkanes.* The Journal of Physical Chemistry B, 1999. **103**(21): p. 4508-4517.

10. Laso, M., J.J.d. Pablo, and U.W. Suter, *Simulation of phase equilibria for chain molecules.* The Journal of Chemical Physics, 1992. **97**(4): p. 2817-2819.

11. Mazzeo, M.D., M. Ricci, and C. Zannoni, *The Linked Neighbour List (LNL) method for fast off-lattice Monte Carlo simulations of fluids.* Computer Physics Communications, 2010. **181**(3): p. 569-581.

(214x) Ionic Liquid Design By Multivariate Characterization and Modeling

Hada, S., Auburn University
Davis, S., Auburn University
Herring, R. III, Auburn University
Eden, M. R., Auburn University

Product design is now considered as an emerging paradigm in the field of chemical engineering because it requires a different set of tools and skill sets from other problems traditionally encountered in the field. Design of environmentally benign solvents and alternative media for extraction and purification are new challenges within product design. One class of novel compounds being studied for such application is ionic liquids (ILs). Ionic liquids that have tailored structures with an array of unique functional properties can have important applications in areas such as CO_2 capture and sequestration, sulfur removal from fuels, energy storage, biomass pre-treatment, and chemical separations. Through variation of both cation and anion, particular ionic liquids with tunable physical properties can be tailored.

However, it is estimated that over 10^{14} unique cation/anion combination are possible for use as room temperature ionic liquids, the majority of which have never been synthesized [[i]]. However, the traditional experimental trial-and-error approach of searching through this large molecular space is unrealistic as it is both time and labor intensive. Thus, it is essential to develop a logical and systematic approach of selectively choosing a given ionic pair that matches a set of desired physico-chemical property targets. The product design framework developed in this work seeks to address this issue by systematically selecting and designing ionic liquid molecules that contain the correct combination of cation and anion that match a set of target physico-chemical properties of interest.

To meet this objective, infrared (IR) spectra of a representative training set of ionic liquids that contain large quantities of descriptor data involving information on molecular architecture is generated using density function theory (DFT). DFT based simulation make the characterization techniques independent of the availability of experimental spectroscopic data. To identify systematic patterns and important features of the molecular architecture in such multivariate data, multivariate statistical techniques such as principal component analysis (PCA) and partial least square (PLS) are used. Appropriate latent variable property models are developed to capture the underlying relationships between physical-chemical properties and/or molecular architecture. Since, conventional regression based GCM does not have property parameters for groups/fragments of ionic molecules, the characterization based group contribution method (cGCM) is utilized for property estimations. The reverse design of potential IL molecules is accomplished by exhaustively generating combinatorial structures from the set of basic groups, which represent the chemical make-up of the training set fragments, until the resulting properties match the target property values. In exhaustive searches, selection from among numerous permutations of anion, cation, and alkyl chain attached to cation groups is performed [[ii]].

[[i]] Turner, E. A., Pye, C. C., & Singer, R. D. (2003). Use of ab Initio Calculations toward the Rational Design of Room Temperature Ionic Liquids. J. Phys. Chem., (Table 1), 2277–2288.

[[ii]] Hada, S., Herring III, R., Eden, M. (2013). Design of Ionic Liquids Using Property Clustering and Decomposition Techniques. Computer-Aided Chemical Engineering, 31, In Press.

(214w) Simulation of Molecules of Arbitrary Geometry Using GO-MC (GPU-Optimized Monte Carlo)

Mick, J. R., Wayne State University
Rushaidat, K. I., Wayne State University
Li, Y., Wayne State University
Hailat, E., Wayne State University
Schwiebert, L., Wayne State University
Potoff, J. J., Wayne State University

The simulation of systems containing over 100,000 atoms has become commonplace with the development open-source parallelized molecular dynamics simulation codes, such as LAMMPS and NAMD [1, 2]. In addition to outstanding scaling across CPUs, these codes, and others, such as HOOMD [3], have been written to utilize inexpensive graphics processors. For the simulation of processes at constant temperature and or pressure, molecular dynamics is clearly the method of choice, however, chemical or biological processes that occur at constant chemical potential require simulation in an ensemble that allows for fluctuations in the number of molecules. Simulation in open systems is performed typically with Monte Carlo simulations in either the grand canonical or Gibbs ensembles. However, the large system sizes [4, 5] required for biomolecular systems make such calculations computationally prohibitive using existing serial codes.

In this work, a multi-ensemble Monte Carlo simulation engine, GO-MC(GPU-Optimized Monte Carlo), capable of simulating systems of tens of thousands of molecules of arbitrary geometry on a desktop workstation is presented. The engine is built on past code to simulate linear molecules on the GPU [6, 7] and is written in C++ using NVIDIA Corp.'s Compute Unified Device Architecture (CUDA) API. Performance comparisons are made between GPU implementations of the coupled-decoupled and a coupled configurational bias methods [8-10]; as measured by acceptance rates and computation time on the device. The coupled approach, while inefficient on the CPU, shows promise on the GPU due to better thread balancing. Performance improvements due to the use of a nearest neighbor cell list [11] are discussed. Further code optimizations available on the NVIDIA Kepler architecture are described.

1. Brown, W.M., et al., *Implementing molecular dynamics on hybrid high performance computers - short range forces.* Computer Physics Communications, 2011. **182**(4): p. 898-911.

2. Phillips, J.C., J.E. Stone, and K. Schulten, *Adapting a message-driven parallel application to GPU-accelerated clusters*, in *Proceedings of the 2008 ACM/IEEE conference on Supercomputing*2008, IEEE Press: Austin, Texas. p. 1-9.

3. Nguyen, T.D., et al., *Rigid body constraints realized in massively-parallel molecular dynamics on*

graphics processing units. Computer Physics Communications, 2011. **182**(11): p. 2307-2313.

4. Freddolino, P.L., et al., *Molecular Dynamics Simulations of the Complete Satellite Tobacco Mosaic Virus.* Structure (London, England : 1993), 2006. **14**(3): p. 437-449.

5. Sanbonmatsu, K.Y., S. Joseph, and C.-S. Tung, *Simulating movement of tRNA into the ribosome during decoding.* Proceedings of the National Academy of Sciences of the United States of America, 2005. **102**(44): p. 15854-15859.

6. Jason R. Mick, K.R., Eyad Hailat, Vincent Russo, Loren Schwiebert, Jeffrey J. Potoff, *GPU Accelerated Configurational Bias Monte Carlo Simulations of Linear Alkanes.* Proceedings of the 2013 AIChE Annual Meeting, 2012: p. 51e.

7. Jason R. Mick, K.R., Eyad Hailat, Vincent Russo, Loren Schwiebert, Jeffrey J. Potoff, *GPU-accelerated Gibbs ensemble Monte Carlo simulations of Lennard-Jonesium"* Comp. Phys. Comm. (submitted, in review), 2013.

8. Frenkel, D. and B. Smit, *Understanding molecular simulation: from algorithms to applications*2002: Academic Press.

9. Martin, M.G. and J.I. Siepmann, *Novel Configurational-Bias Monte Carlo Method for Branched Molecules. Transferable Potentials for Phase Equilibria. 2. United-Atom Description of Branched Alkanes.* The Journal of Physical Chemistry B, 1999. **103**(21): p. 4508-4517.

10. Laso, M., J.J.d. Pablo, and U.W. Suter, *Simulation of phase equilibria for chain molecules.* The Journal of Chemical Physics, 1992. **97**(4): p. 2817-2819.

11. Mazzeo, M.D., M. Ricci, and C. Zannoni, *The Linked Neighbour List (LNL) method for fast off-lattice Monte Carlo simulations of fluids.* Computer Physics Communications, 2010. **181**(3): p. 569-581.

(214z) A Molecular Dynamics Study of the Effects of Ionic Liquids On Human Serum Albumin

Jaeger, V., University of Washington
Pfaendtner, J., University of Washington

Ionic liquids (IL), which are organic salts that are liquid below 100° C, have been studied as a unique solvent for biologically relevant molecules including proteins. Research has shown that ionic liquids hold promise as a biological solvent in a few cases. First, they solvate recalcitrant solutes like cellulosic biomass. Second, they can affect the function of enzymes in order to bring about unique reactions. Third, they can stabilize proteins during long-term storage or facilitate refolding of proteins. One class of model protein that has been studied for its stability in ILs is serum albumin, a ubiquitous protein that facilitates the transport of molecules through the bloodstream. These studies have provided insights into the effects of ionic liquids on the function, structure, and dynamics of serum albumin, but most studies lack a molecular level understanding of the protein-solvent interaction. We have employed molecular dynamics with enhanced sampling techniques in order to compliment the breadth of published work and to gain molecular level insight into this complex interaction.

We conducted molecular dynamics simulations to test the effects of imidazolium- and choline-based ILs in binary mixtures with water on the structure and dynamics of human serum albumin. The dynamic fluctuations of the protein are dampened by the presence of IL as measured by root mean squared fluctuation. Principle component analysis shows differences in the slow modes of motion depending on the composition of the solvent. Serum albumin contains several well defined domains that are held together by disulfide bridges that cause the protein to adopt a heart-shaped conformation. Essential dynamics coarse graining was employed to determine if the location of each domain remains the same in each solvent. Structuring of the IL and water around the protein was analyzed to determine preferential interaction sites and the effects of these interactions. All of the results show qualitative agreement with experimental observations.

One specific study indicates that loop 1 of domain I in human serum albumin displays high rates of denaturation in imidazolium-based ILs as measured by fluorescence spectroscopy. We employ well-tempered metadynamics in order to study this specific loop. Well-tempered metadynamics is an enhanced sampling technique in which a small time-dependent bias potential is added to a collective variable in order to discourage the system from revisiting previously explored conformations. Solvent dependent differences in the free energy of the transition of this loop from its folded to unfolded state are observed.

(214y) Mechanical Deformation of dsDNA: Molecular Modeling Approach

Hwang, H. H., Georgia Institute of Technology
Chun, B. J., Georgia Institute of Technology
Jang, S. S., Georgia Institute of Technology
Kim, H. D., Georgia Institution of Technology

Mechanical Deformation of dsDNA: Molecular Modeling Approach

Hyea Hennim Hwang, Byeong Jae Chun and Seung Soon Jang

School of Materials Science and Engineering, Georgia Institute of Technology,

771 Ferst Drive, Atlanta, GA 30332-0245

Harold D. Kim

School of Physics, Georgia Institute of Technology,

837 State Street, Atlanta, GA 30332-0430

Studying sequence-dependent DNA dynamics is crucial for understanding the relationship between DNA sequence and genome function, which is one of the holy grails of biology. For this purpose, molecular dynamics (MD) simulation can provide detailed information on structural changes of DNA at resolution not yet accessible to current experimental techniques. In this study, we use MD simulation to investigate the structural changes that double-stranded DNA (dsDNA) undergoes during mechanical deformation and measure how stress is distributed in this process. In our simulations, a 9 base pair B-DNA duplex is tethered at two H-terminated Si (111) surfaces through spacers, and is extended by pulling these surfaces apart at a constant rate. The change of stress required for extension is monitored as a function of strain. During the extension, the structural change is analyzed in terms of the hydrogen bonding geometry in each base pair, the torsion angles along backbone, diameter of double helical structure. Transition state from double helix to a novel ladder structure is also analyzed, in which base pairs of one strand are stacked on those of the other strand. By designing the base sequence for simulations, the sequence-dependency of mechanical deformation is elucidated.

(214aa) Ab Initio Modeling of the Direct Interaction of Surface Plasmons With Adsorbates On Metal Nanostructures

Morabito, M., University of Michigan
Xin, H., University of Michigan
Linic, S., University of Michigan

We recently reported on a new family of photo-catalysts that can efficiently couple thermal energy and photonic stimuli to drive photo-chemical transformations with unprecedentedly high efficiencies.[1,2,3] These materials are plasmonic nanostructures of coinage metals which respond to low intensity visible light through a formation of a collective, oscillating electrons.

We will discuss our approach to study these reactions from first principles. We have employed Langevin dynamics simulations to investigate the activation of adsorbed diatomic molecules over plasmonic metal nanostructures.[4] Using these simulations we calculated molecular dissociation probabilities and reaction rates at different light intensities. These stochastic simulations combine Newton's equations of motion with electronic drag from the surface. The electronic friction is caused by partial charge transfer between substrate and adsorbed molecule due to changes in the electronic structure induced by direct interactions of adsorbates with low energy electron in plasmons. We calculated electronic friction from the adiabatic electronic wavefunctions from Density Functional Theory, using a quantum model to obtain an electron distribution in the surface plasmon, a collection of low-energy excited electrons.[5][6] We compare this approach to desorption induced by (multiple) electronic transitions (DIET/DIMET) model, another mechanism proposed for this type of plasmonic photocatalytic enhancement.

[1] P. Christopher, H. Xin, M. Andiappan and S. Linic. Nat. Mat. **11**, 1044 2012

[2] P. Christopher, H. Xin, S. Linic. Nat. Chem. **3**, 467 2011

[3] S. Linic, P. Christopher, D. B. Ingram. Nat. Mat. **10**, 911 2011

[4] T. Olsen and J. Schiøtz. J. Chem. Phys. **133**, 034115 2010

[5] V. Krishna and J. C. Tully. J. Chem. Phys. **125**, 054706 2006

[6] L. Genzel, T. P. Martin, U. Kreibig. Z. Physik B. **21**, 339 1975

(214ab) Molecular Modeling and Characterization of Dextran-Coated Nanoparticles in Aqueous Medium

Wang, C., Missouri University of Science and Technology
Wang, J. C., Missouri University of Science and Technology

Dextran is an important type of polysaccharide that has good biocompatibility, biodegradability, aqueous solubility, and many hydroxyl (-OH) groups for further functionalization. It has also been increasingly used in bio-related applications. Dextran-coated nanoparticles can represent a material system that provides the benefits of both dextran and nanoparticle simultaneously to further advance the field of nanobiotechnology. The grafted dextran chains on the nanoparticle surface form a porous layer whose pore structure greatly determines the properties and functions of the nanoparticle but has yet to be characterized. In this work, the molecular dynamics (MD) methodology is employed to model and simulate dextran-coated solid nanoparticles in an aqueous environment. A combination of simulation snapshots, dual grid-based pore structure analysis, and radial distributions is use to achieve a quantitative and detailed characterization of the pore structure formed by the dextran chains grafted at a high surface density. The results show that most of the dextran chains are able to retain highly helical conformations due to their limited structural flexibility from relatively short chain length and additional support from conformationally frozen monomers and a high chain density in the base region. Although most pore openings are in the smallest size range, in the outer region significantly larger pores exist but become increasingly smaller when the pores extend towards the nanocore. Water molecules are found to be pervasive throughout the dextran layer and the nanocore surface, indicating good solvation of the dextran-coated nanoparticle by water

(214ac) Molecular Simulation Studies of Adsorption and Diffusion of Complex Molecules in Zeolites

Bai, P., University of Minnesota
Tsapatsis, M., University of Minnesota
Siepmann, J. I., University of Minnesota

Zeolites are very important materials both in the industrial world and to our everyday life. Numerous studies have probed the adsorption mechanisms of zeolites with the goal of designing sorbents with higher capacity and selectivity, and computer simulations have traditionally been applied to gain molecular-level insights for these adsorption systems. However, most simulation studies are limited to small guest molecules without strongly directional interactions, such as short alkanes or CO_2, while more complex molecules are challenging because of a lack of transferable force fields and efficient algorithms. We present here the development of the TraPPE-zeo force field, that enables the accurate modeling of the adsorption of any type of molecule in all-silica zeolites, and the use of various configurational-bias Monte Carlo moves for enhanced sampling. These methodological advances allow one to tackle problems of practical interest, such as the adsorption of alcohols, polyols, and furfurals from aqueous solution and of long-chain alkanes from liquid mixtures.

(214ad) Ab Initio Excitation Energy Transfer in Peridinin-Chlorophyll a-Protein Complex

Bricker, W. P., Washington University in St. Louis
Lo, C. S., Washington University in St. Louis

The study of excitation energy transfer in photosynthetic light-harvesting antenna complexes is important because of their high quantum efficiencies, which can approach 100% between pigments in some systems. Peridinin-Chlorophyll *a*-Protein (PCP) is a light-harvesting complex which has been isolated from the dinoflagellate *Amphidinium carterae* and characterized by x-ray crystallography. PCP is a trimeric light-harvesting protein complex which contains eight peridinin carotenoids and two chlorophyll *a* pigment molecules per protein monomer, and the excitation energy transfer efficiency within this system is estimated at close to 90%. To study the excitation energy transfer in PCP, we employ the Förster Resonance Energy Transfer (FRET) model, which assumes that energy transfer between pigments occurs by an incoherent dipole-dipole interaction. We calculate the FRET coulombic coupling and spectral overlap terms using *ab initio* multireference configuration interaction (MR-CI) with semi-empirical basis sets. The two dominant excitation energy transfer pathways within PCP are from the S_2 state of peridinin to the Q_x band of chlorophyll a, and from the S_1/ICT (intramolecular charge transfer) state of peridinin to the Q_y band of chlorophyll *a*. In peridinin, absorption in the visible spectrum is due to the strongly allowed $S_0 \rightarrow S_2$ transition, while the $S_0 \rightarrow S_1$ transition is optically forbidden as well as having significant double excitation behavior. While the other transitions in PCP are single excitations, we must model our system using MR-CI to accurately describe the double excitation behavior of the S_1/ICT state in peridinin. Our excitation energy transfer study of PCP using FRET with parameters derived from *ab initio* MR-CI reveals the interplay of pigment geometry and environment on the relative contributions of energy transfer rates and efficiencies within the complex as a whole.

(214am) Computing Effective Bond Orders in Periodic Materials

Manz, T., New Mexico State University

While many methods are available for computing effective bond orders (EBOs) in non-periodic (e.g., molecular) systems, only a few methods are available for computing EBOs in periodic materials. The EBO between two atoms can be generically defined as the number of electrons exchanged between them. Different definitions of the exchange projectors lead to different EBO methods. This talk will describe a new method for computing EBOs in materials with any number (0, 1, 2, or 3) of periodic boundary conditions. The method's primary advantage is computational convenience: (a) it requires only the electron density and spin distribution as inputs, (b) its computational cost scales linearly with increasing unit cell size, and (c) it converges robustly. The method's accuracy is assessed by careful comparisons to systems whose bond orders are inferred from experimental measurements and general chemical properties. These assessments show the method is highly accurate. For molecular systems, results are benchmarked against established methods for computing EBOs to further verify its accuracy and reliability. Results will be presented for a variety of systems including: (a) metallic solids, (b) covalently bonded solids, (c) ionic solids, (d) solid surfaces, and (e) molecular systems. These materials include non-magnetic systems, systems with collinear magnetism, and systems with non-collinear magnetism. Of particular note, the method is shown to be accurate for metallic conductors, which other EBO methods have difficulty describing.

(214ae) Nucleation of Two-Dimensional Boronate Ester-Linked Covalent Organic Framework Moieties in Solution

Koo, B., Cornell University
Clancy, P., Cornell University

Nucleation of Two-Dimensional Boronate Ester-Linked Covalent Organic Framework Moieties in Solution

Brian Koo, Paulette Clancy

Chemical and Biomolecular Engineering, Cornell University

Covalent organic frameworks (COFs) are porous materials that are highly regarded for their potential applications in gas storage, catalysis, and optoelectronics. They rival metal-organic frameworks and zeolites in terms of surface area, weight, and porosity. Predictable assembly of COFs *via* covalent bond-forming reactions from simple building blocks has led to the discovery of many structurally diverse 2D and 3D networks, but the mechanisms governing their nucleation and growth are not understood. Although equilibrium stacking properties of many 2D COF sheets have been experimentally and computationally measured, the effect of solvent, temperature, and surface effects on the thermodynamic processes leading to the growth of thin-film and crystallite COFs has not been studied. COF-forming reactions are reversible in solution and thus lead to the formation of the most thermodynamically stable products.

Our computational approach explores the thermodynamics of nucleation of two COFs, COF-5 and Pc-PBBA, by following their nucleation from individual building blocks (octahydroxyphthalocyanine, hexahydroxytriphenylene, and 1,4-phenylenebis(boronic acid)) in a mesitylene and dioxane solution. We calculate the relative free energies of each boronate ester-linked Pc-PBBA and COF-5 moiety of increasing size to determine likely intermediate structures smaller than the critical nucleus. We then compare the differences between the free energy of growth in the lateral direction (forming larger 2D sheets), and growth in the stacking direction (forming thicker stacks of 2D sheets).

We combine thermodynamic integration and steered molecular dynamics to calculate binding free energies associated with stacking and use entropy calculations to estimate the reaction free energies associated with lateral growth. We follow reversible pathways from individual building blocks to COF fragments with multiple formula units and obtain free energy differences between intermediate products. These studies suggest that lateral growth of COF fragments proceeds more readily at first than vertical stacking, followed by a preference for vertical stacking once COF fragments reach sufficient size. From these results, an estimate of the critical nucleus size for COF growth in solution can be made. By performing free energy calculations of COF moieties we are able to resolve mechanisms of growth too rapid and species too small to be observed experimentally.

(214ah) Molecular Dynamics Simulations of Cellulose Interaction With Polymeric Solid Acid

Sun, X.,
Du, H., University of Arkansas
Wickramasinghe, S. R., University of Arkansas
Qian, X., University of Arkansas

A novel polymeric solid acid catalyst has been designed for cellulose depolymerization. A poly (styrene sulfonic acid) (PSSA) polymer chain is immobilized on substrate surface and used to catalyze biomass hydrolysis. A neighboring poly (vinyl imidazolium chloride) ionic liquid (PIL) polymer chain is grafted from the surface to help solubilize lignocellulosic biomass and enhance the catalytic activity. Over 97% and 32% total reducing sugar (TRS) yields were obtained for cellulose hydrolysis in [EMIM]Cl and water solutions respectively. The catalytic activity for cellulose hydrolysis is found to depend strongly on the chain length and chain density of the two polymeric nanostructures. Further, the interactions between the cellulose and the polymer chains also play a critical role in the conversion rate. Here classical molecular dynamics simulations were conducted to investigate the interactions between the PSSA and PIL polymer chains and the cellulose substrate in both aqueous solution and ionic liquid solvent. The effects of copolymerization and ring substitution on the polymer chains were determined.

(214af) A Simulation Study of Equilibrium States in Supercooled Stillinger - Weber Silicon

Gautam, A. K., Indian Institute of Technology Kanpur
Apte, P. A., Indian Institute of Technology Kanpur

In the supercooled liquids which are prone to crystallization, identification of properties of metastable equilibrium liquids can be challenging. In case of supercooled silicon, modeled by the Stillinger—Weber (SW) potential [F. H. Stillinger and T. A. Weber, Phys. Rev. B. 32, 5262 (1985)], several earlier studies have identified 1060 K as the lowest temperature at which it is possible to equilibrate the liquid at zero pressure. We discuss how the equilibrium liquids at or above 1060 K can be identified based on free energy calculations [P. A. Apte and A. K. Gautam, J. Stat. Phys. 149, 551 (2012)]. In particular the Gibbs free energies of these liquids must satisfy the Gibbs--Helmholtz relation for equilibrium states. Further we also find that the potential energy distributions generated through isothermal isobaric simulation trajectories at various temperatures follow the equilibrium temperature—potential energy curve closely. We also discuss some interesting aspects of relaxation of the supercooled liquids vis-a-vis the properties of the equilibrium states.

(239a) Theoretical Investigation of Ethylene Glycol Reforming Over Pt(111) in Vapor and Aqueous Environments

Faheem, M., University of South Carolina
Heyden, A., University of South Carolina
Originally presented on: 11/5/2013 8:30:00 - 8:45:00

Catalytic conversion of biomass-derived oxygenates for production of hydrogen and transportation fuels has gained significant attention in recent years. Ethylene glycol has been used as a model molecule to gain mechanistic insights into the reaction mechanism. The main reaction pathway for the production of hydrogen involves cleavage of C-C, C-H, and O-H bonds with subsequent water-gas shift reaction. A parallel pathway leads to the production of alkanes through C-O bond scissions. While the reaction mechanism is to some degree understood at the metal-gas interface, lack of a well-established methodology for describing the influence of a complex liquid on a reaction across a solid-liquid interface with quantum mechanical accuracy has resulted in very little knowledge on the effect of a liquid-phase environment on the reaction mechanism.

In this talk, we present an application of our recently developed iSMS[1] methodology to study the mechanism of ethylene glycol reforming over Pt(111) in water. Periodic planewave density functional theory calculations have been performed to obtain optimized structures for all intermediates and to calculate free energies of reaction and activation free energy barriers in the vapor phase. The effect of water has then been described through a combination of Gaussian-type orbital calculations and COSMO-RS calculations as required by iSMS. Finally, detailed microkinetic models have been developed for both vapor and aqueous phases to provide insights into the effect of a liquid phase environment on the reaction mechanism and activity descriptors.

1. Faheem, M.; Suthirakun, S.; Heyden, A. *J. Phys. Chem. C* **2012,** *116*, 22458-22462.

(239b) Designing Catalysts Using Predictions of Hydrocarbon and Oxygenate Adsorption Energies

Montemore, M. M.,
Medlin, J. W., University of Colorado Boulder
Originally presented on: 11/5/2013 8:45:00 - 9:00:00

Oxygenates and hydrocarbons are important in most catalytic processes, such as alcohol oxidation, biomass conversion, petrochemical production, fuel cells, Fischer-Tropsch synthesis, and CO_2 electroreduction. The heterogeneous catalysts used in these processes are generally found through a trial-and-error process. Given the immense combinatorial space for multimetallic surfaces, more efficient design paradigms are needed. Adsorption energies have been shown to be good predictors of catalytic performance; hence, efficient predictions of adsorption energies could greatly improve catalytic design.

We have developed an accurate model for predicting adsorption energies of hydroxyl and methyl in the top site from electronic structure parameters. This model uses the d-band center, p-band center, matrix coupling element, d-band filling, and electronegativity of the surface. These quantities are readily available from density functional theory (DFT) calculations, and can also be estimated directly from alloy structures. Simple relations based on metallic parameters can also be used to predict the energy of translating these species to more highly coordinated sites. Hence, these relations allow very efficient screening of surfaces for their adsorption properties.

We have also developed accurate scaling relations that allow prediction of adsorption energies of other oxygenates and hydrocarbons from the adsorption energies of hydroxyl and methyl. For example, the adsorption energy of longer alkyls can be predicted using methyl's adsorption energy as well as gas-phase bond energies. Previous scaling relations among C_1 hydrocarbons have been made more accurate by scaling between these adsorbates in the sites where the tetravalency principle is fulfilled. These relations allow efficient prediction of potential energy surfaces for a variety of reaction pathways based on adsorption energies calculated from electronic structure parameters or calculated using DFT.

Together, these relations allow the rational design of a catalyst based on electronic structure parameters, as well as efficient screening from DFT calculations. Hence, we have carried out a comprehensive search for surfaces that catalyze alcohol oxidation, the oxygen reduction reaction, and alkane reforming.

(239c) Effect of Palladium Surface Structure On the Hydrodeoxygenation of Organic Acids

Behtash, S., University of South Carolina
Lu, J., University of South Carolina
Heyden, A., University of South Carolina
Originally presented on: 11/5/2013 9:00:00 - 9:15:00

Lipid feedstock can be utilized for the production of second-generation biofuels via a catalytic hydrodeoxygenation (HDO) process. The conversion of fatty acids plays an important role in the activity and selectivity of this process. Understanding the HDO reaction mechanism of organic acids on metal surfaces is a prerequisite for the rational design of new HDO catalysts specifically designed for lipid feedstock. In our previous study of the HDO of propionic acid over Pd (111), we found that the dehydrogenation of α-carbon and the C-OH bond dissociation steps are activity descriptors in the HDO of organic acids. However, it is well known that the higher-lying d-states in under-coordinated stepped surfaces lower the C-O and C-C bond dissociation barriers and consequently enhance the reactivity in comparison to flat surfaces. To investigate the sensitivity of the HDO of organic acids on the Pd surface structure, the reaction mechanism of the HDO of propionic acid has been studied on the Pd(211) model surface and the results are compared to our previous Pd(111) investigation. Periodic density functional theory (DFT) calculations have been carried out to obtain reaction rate parameters which are subsequently used in a microkinetic modeling study. Finally, we performed a sensitivity analysis and computed Campbell's degree of rate control to identify rate-determining steps.

(239d) Computational Design of Copper Containing Bimetallic Syn-Gas to Ethanol Catalysts

Bruce, D. A., Clemson University
He, M., Clemson University
Originally presented on: 11/5/2013 9:15:00 - 9:30:00

A major challenge associated with the synthesis of ethanol from syn-gas is an inability to find a low-cost catalyst that promotes the proper combination of CO dissociation and CO insertion steps, so as to yield ethanol as the primary reaction product and inhibit the formation of methane, methanol, longer chain alkanes, and other coking reaction products. Given the complexity of this reaction system, a simple trial-and-error approach to catalyst design is fraught with difficulties, which could severely limit efforts to identify an ideal catalyst material. Therefore, we used quantum mechanical simulations for the rational design of bimetallic catalysts that are optimally suited for the production of ethanol from syn-gas. Specifically, Density Functional Theory (DFT) simulations and Bronsted-Evans-Polanyi (BEP) relations were used to map out the full reaction mechanism from syn-gas to ethanol for 4 different copper containing bimetallic clusters (CuCo, CuNi, CuFe, and CuRu) that range in size from 13 to 38 metal atoms. For these bimetallic catalysts, microkinetic models were built, considering all necessary adsorption and reaction steps as well as the diffusion of intermediate species between varying metal surface sites. Results from additional computational studies examining surface coverage effects were also included in the microkinetic models. These simulations indicate the nature and stability of the various bimetallic nanocatalysts and more importantly identify specific metal combinations that are ideally suited for ethanol production. Select catalyst formulations (e.g., CuCO and CuNi) were prepared and tested for catalytic activity and selectivity. These experimental results were used to validate the computational results.

(239e) Improved Catalytic Reactivity of PdFe Bimetallic Catalysts Towards the Deoxygenation of Phenolic Compounds: A Density Functional Theory Study

Hensley, A., Washington State University
Zhang, R., Washington State University
Wang, Y., Pacific Northwest National Lab
McEwen, J. S., Washington State University
Originally presented on: 11/5/2013 9:30:00 - 9:45:00

With an increasing world population along with the emergence of newly industrialized nations, the availability of fossil fuels will significantly decrease as current oil wells and fields are depleted. One of the most promising renewable energy alternatives to fossil fuels are biofuels because of their compositional similarity to fossil fuels and their ability to use the current transportation fuel infrastructure[1]. The major problem facing biofuels is the high level of oxygen, up to 40 wt%, present in the crude bio-oil after pyrolysis which cause the resulting fuel to be corrosive and thermally unstable[2]. To address this problem, a novel PdFe catalyst was recently developed for the purpose of removing the oxygen functional groups from the phenolic compounds found in crude bio-oil[3]. Experimentally, this bimetallic catalyst showed a significant catalytic improvement for the hydrodeoxygenation (HDO) of guaiacol to benzene over its monometallic components[4]. To gain a deeper understanding of the interactions occurring within the PdFe bimetallic catalyst and their effect on the HDO of phenolic compounds, the adsorption of benzene, phenol, and guaiacol on model Fe (110), Pd (111), and mixed PdFe surfaces along with the HDO reaction path for the conversion of phenol to benzene on the monometallic surfaces were examined using density functional theory (DFT). Our results show that while the adsorption strength for these aromatic species on the pure surfaces was nearly identical, the Fe surface distorted the C-O bonds to a greater degree than the Pd surface. When these compounds were adsorbed onto mixed PdFe surfaces, their adsorption was found to be weakened by the interaction with the surface Pd. In particular, we find that benzene is physisorbed when one monolayer of Pd is deposited on Fe(110) while it is chemisorbed on a pure Fe(110) surface.[5] In addition to the adsorption studies, the reaction pathway and energy barriers for the HDO of phenol to benzene on the monometallic model surfaces were calculated. Preliminary results show that the energy barrier for the HDO reaction on the pure Fe surface is significantly reduced as compared to the pure Pd surface. Overall, these results strongly support the hypothesis that the Fe surface acts as the active catalytic surface.

References

1. Lovins, A. B.; Datta, E. K.; Bustnes, O.-E.; Koomey, J. G.; Glasgow, N. J., *Winning the Oil Endgame: Innovation for Profits, Jobs, and Security*. Rocky Mountain Institute: 2005.

2. Mohan, D.; Pittman, C. U.; Steele, P. H., Pyrolysis of Wood/Biomass for Bio-oil: A Critical Review. *Energy & Fuels* 2006**,** *20*, 848-889.

3. Sun, J.; Karim, A. M.; Zhang, H.; Kovarik, L. S.; Wei, Z.; Li, X.; Wang, Y. In *Hydrodeoxygenation of Biomass-Derived Compounds to Biofuels*, AIChE Annual Meeting, Pittsburgh, PA, Pittsburgh, PA, 2012.

4. Sun, J.; Karim, A. M.; Zhang, H.; Kovarik, L.; Li, X.; Hensley, A. J.; McEwen, J.-S.; Wang, Y., Carbon Supported Bimetallic Pd-Fe Catalysts for Vapor-Phase Hydrodeoxygenation of Guaiacol. Washington State University: Journal of Catalysis (submitted), 2013.

5. Hensley, A. J.; Zhang, R.; Wang, Y.; McEwen, J.-S., Energetic and Electronic Interactions between Benzene and PdFe Surfaces: A Density Functional Theory Study. Washington State University: (in preparation), 2013.

(239f) Multiscale Kmc-DFT Simulation On Bimetallic Catalysts
Guo, W., UNIVERSITY OF DELAWARE
Originally presented on: 11/5/2013 9:45:00 - 10:00:00

Multiscale KMC-DFT Simulation on Bimetallic Catalysts

Wei Guo[1], Michail Stamatakis[1,2] and Dionisios G. Vlachos[1]*

[1]Center for Catalytic Science and Technology, Department of Chemical and Biomolecular Engineering, University of Delaware, Newark, Delaware 19716, USA

[2]Department of Chemical Engineering, University College London, Torrington Place, London WC1E 7JE, UK

* vlachos@udel.edu

Bimetallic catalysts have provided an opportunity for designing more active, stable and less expensive materials than the individual, pure metals catalysts. However, the reaction mechanism and the dynamics of the catalyst during catalytic reaction are not yet fully understood. The effect of fine microstructure of Ni clusters on catalyst reactivity remains elusive. Here we have developed a multiscale approach to simulate the nitrogen desorption from Ni clusters of different sizes and shapes on Pt(111) (Ni/Pt). We employed density functional theory (DFT) calculations to model the coverage dependent N binding energies on a variety of Ni/Pt configurations, and presented a model based on local coordination environment of Ni neighbors around a nitrogen atom [1]. Then the energetics of adsorption, surface diffusion, association and desorption of atomic N on Ni/Pt catalysts are incorporated into a recently developed kinetic Monte Carlo (KMC) framework to simulate the temperature-programed desorption (TPD) spectra of nitrogen on Ni /Pt [2]. This is the first study of first-principles KMC simulation on bimetallic catalysts that provides a conceptual platform for accounting for complex microstructure on kinetics. Our simulation framework can be employed to understand the structure sensitivity in catalysis and eventually tune bimetallic catalytic materials for complex chemistries.

1. W. Guo and D. G. Vlachos, J Chem Phys **138**(17), 174702 (2013).

2. W. Guo, M. Stamatakis and D. G. Vlachos, Submitted (2013).

(239g) Investigation of Hydrogen Spillover Pathways On a Bimetallic Hydrogenation Catalyst

Ruscillo, F., University College London
Sykes, C., Tufts University
Stamatakis, M., University of Delaware
Originally presented on: 11/5/2013 10:00:00 - 10:15:00

Hydrogenation processes are of paramount importance in the chemical industry with a vast range of applications in the food, petrochemical and pharmaceutical sectors. Heterogeneous catalysts are widely used for hydrogenation; however, controlling the spillover of hydrogen on catalytic surfaces is challenging. It has recently been shown experimentally that it is possible to achieve this type of control on a Cu/Pd catalyst by reversible adsorption of a spectator molecule (CO).[1] This phenomenon was termed the "molecular cork" effect and can have profound implications in catalysis. In particular, it is possible to trap hydrogen on the catalytic surface at temperatures higher than its normal desorption temperature, thereby increasing the rate of hydrogenation reactions due to the higher reactivity of hydrogen under these conditions. Yet, questions remain with regard to the origins of this phenomenon and how it can be leveraged for the design of improved hydrogenation catalysts.

Motivated by these questions we employ a multiscale computational approach to study the "molecular cork" system. At the molecular level we use density functional theory (DFT) to calculate the energetics and kinetics of CO adsorption onto and desorption from the Pd sites, as well as hydrogen dissociative adsorption and associative desorption in the absence and in the presence of CO. At the catalyst level, we use kinetic Monte Carlo simulation to reproduce the experimental observations and investigate in detail the H_2 dissociation and spillover pathways. Our simulations provide valuable information that cannot be directly accessed by experiments and offer a fundamental understanding of the molecular-level processes underlying the "molecular cork" effect.

References

Marcinkowski, M. D., A. D. Jewell, M. Stamatakis, M. B. Boucher, E. A. Lewis, C. J. Murphy, G. Kyriakou and E. C. H. Sykes, *Controlling a Spillover Pathway with the Molecular Cork Effect.* Nature Materials, 2013: Accepted for publication.

(239h) Understanding the Alcohol Dehydration On Metal Oxides

Mpourmpakis, G., University of Delaware

Originally presented on: 11/5/2013 10:15:00 - 10:30:00

A fundamental understanding on the dehydration of simple alcohols can help us elucidate and eventually control the selective dehydration of polyols and complicated biomass molecules to value-added chemicals. Herein, we use Density Functional Theory (DFT) calculations to study the alcohol dehydration on group XIII oxides, including γ-Al_2O_3 which is a commonly used support in heterogeneous catalysis. Both E1 and E2 type of dehydration mechanisms for primary (ethanol, 1-propanol), secondary (2-propanol), and tertiary (2-methyl-2-propanol) alcohols were investigated. In addition, we considered possible coupling reactions between reactants, products and dehydration intermediates. A key finding is that the dehydration barriers of the alcohols correlate linearly with the corresponding carbenium ion stability (alcohol property) [1]. This is because the transition state of the alcohol dehydration has carbenium characteristics. By accounting for the acid strength of the active site, the basic strength of the surface oxygens and the substitution degree of the alcohol, structure-property relationships were developed. Our results explain several experimental observations.

[1] Roy S., Mpourmpakis G., Hong D.-Y., Vlachos D.G., Bhan A. and Gorte R.J., *ACS Catal.* **2**, 1846–1853 (2012)

(239i) CeO2-Supported Ni Catalysts For CO2 Methanation

Hahn, K., University of Zurich
Hutter, J., University of Zurich
Originally presented on: 11/5/2013 10:30:00 - 10:45:00

Combined metal-metal oxide systems are important materials for numerous technological applications such as microelectronics, photovoltaics, gas sensors and in particular in oxide- supported metal catalysts [1]. Deposition of small metal particles on metal oxides such as Al_2O_3, CeO_2 and TiO_2 has led to enhanced catalytic activity, for example, for the water gas shift reaction or CO oxidation [2,3]. Metal oxide supported materials are also promising catalysts for hydrocarbon production from CO_2, a reaction process interesting for alternative energy production since it provides high energy density fuel from CO_2 which is omnipresent in our environment. In fact, CeO_2-supported Ni particles have proven to be promising materials for the activation of methanation of CO2 [4].

Here, density functional theory within the Gaussian and plane waves (GPW) implemented in the Quickstep [5] module of the CP2K [6] program package has been used for the simulation of Ni cluster formation on CeO_2(111) surfaces. Stabilization of Ni particles has been observed with increasing cluster size up to ten atoms. It has been found that O diffusion from the subsurface layer of CeO_2(111) to the surface layer is facilitated in the presence of Ni particles on the surface. Furthermore, the adsorption and dissociation of CO_2 has been investigated on clean CeO_2(111), on non-supported Ni clusters in the gas phase and on the Ni/CeO_2(111) system. This study gives fundamental insight into the metal-metal oxide interactions and its function for CO_2 catalysis.

[1] C.T. Campbell, Surf. Sci. Rep. 27 (1997), 1.
[2] Y. Zhai, D. Pierre, R. Si, W. Deng, P. Ferrin, A. U. Nilekar, G. Peng, J. A. Herron, D. C. Bell, H. Saltsburg, M. Mavrikakis and M. Flytzani-Stephanopoulos, Science 329 (2010), 1633.
[3] G. N. Vayssilov, Y. Lykhach, A. Migani, T. Staudt, G. P. Petrova, N. Tsud, T. Skala, A. Bruix, F. Illas, K. C. Prince, V. Matolin, K. M. Neyman and J. Libuda, Nat. Mater. 10 (2011), 310.
[4] W. Wang and J. Gong, Front. Chem. Sci. Eng. 5 (2011), 2.
[5] G. Lippert, J. Hutter and M. Parrinello, Mol. Phys. 92 (1997), 477.
[6] J. VandeVondele, M. Krack, F. Mohamed, M. Parrinello, T. Chassaing and J. Hutter, Comput. Phys. Commun., 167 (2005), 103.

(239j) Effects of Doped N On the Reaction of Alkyne Hydrochlorination Catalyzed By Graphene Supported Au Based Catalyst

Du, Q., Tianjin University
He, Z.,
Zhang, J., Tianjin University
Li, W., Tianjin University
Han, Y., Tianjin University
Originally presented on: 11/5/2013 10:45:00 - 11:00:00

Many studies have shown that the support can greatly affect the distribution and stability of the catalyst through the interfacial interaction. It is a trend to modify the support to improve the catalytic performance. The aim of our work is to reveal the effect mechanism of the support modification by N-doping on the reaction of alkyne hydrochlorination catalyzed by graphene supported Au based catalyst. This investigation will provide a theoretical basis of how to select optimum N-doped location and valence state to interact with the catalyst and improve the catalytic performance. And it is instructive to the experimental design of novel non-mercury catalysts with high performance.

The graphene edge has two types: zigzag-edge and armchair-edge. Herein, four different N-doped sites are considered in this study: graphitic N, pyridine N on zigzag edge(zigzag 6-N), pyrrolic(armchair 5-N) and pyridine N(armchair 6-N) on armchair. Our results are divided into three sections:**(a) The effects of doped N and its location on the stability of Au based catalyst. (b) The effects of doped N and its location on the adsorption of reactants and product. (c) The effects of doped N and its location on the reaction paths and barriers.**

(269a) A Product Engineering Approach: Diffusion and Performance of Fragranced Products

Teixeira, M. A.,
Rodrigues, A. E., LSRE - Laboratory of Separation and Reaction Engineering - Faculty of Engineering - University of Porto
Originally presented on: 11/5/2013 8:30:00 - 8:55:00

Product Engineering Approach for the Design of High Performance Fragranced Products

Miguel A. Teixeira and Alirio E. Rodrigues*

LSRE -- Laboratory of Separation and Reaction Engineering, Associate Laboratory, Dept. of Chemical Engineering, Faculty of Engineering of University of Porto

Rua Dr. Roberto Frias, 4200-465 Porto, Portugal

Abstract

Chemical Product Engineering has emerged over the last decade as a new paradigm on Chemical Engineering.[1, 2] In fact, this is an answer to a global change in the chemical industry economy: in the past driven by technological expansion for the large production of commodities, towards a market-driven production of high added-value, specialty and performance products as happens today.[2, 3] The relevance of introducing perception, and namely consumer perception, in the design of chemical products has been recently highlighted as an important step for product design. Among the different types of perception, those related to chemical perception (chemosensory) and namely taste and olfaction are still difficult to work with. This is due to the complexity of the process of olfaction at the biochemical level.

Fragrances are presented in countless products we deal with everyday like perfumes, shampoos and soaps among many others. In fact, more than three-quarters of toiletries and household products contain a fragrance in its composition. However, the design of fragranced products is still a combination of art, technology, and some scientific knowledge, although the former predominates at the industry.[4, 5]

In order to reduce subjectivity and empiricism in this field, we have developed an integrated theoretical model to assess the performance of fragrances considering the release and diffusion processes together with the predicted odor intensity over time and distance from the source. In order to achieve that goal, our model combines different scientific fields: Thermodynamics, Transport Phenomena, and Psychophysics into a common product - fragrances. This approach is within Product Design or Engineering, which in this case is applied to Perfume Engineering as shown in Figure 1.

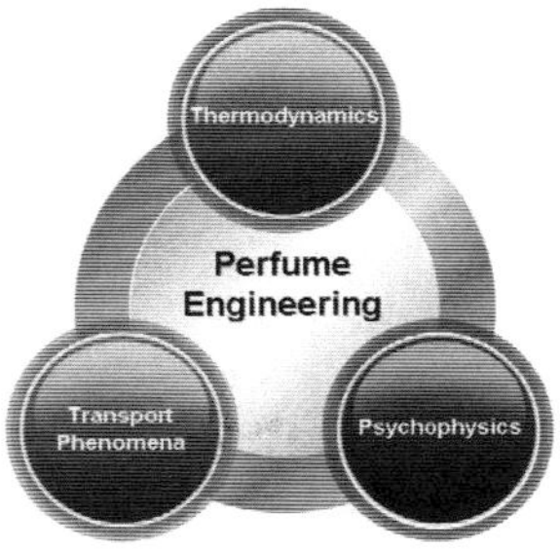

Figure 1 - Our vision of what is Perfume Engineering: the integration of concepts from different fields (Thermodynamics, Transport Phenomena, and Psychophysics) in order to improve the design and development of new perfumed products.[4]

Within this line of though, we can simplify fragranced products to a mixture of *N* fragrance raw materials diluted in a suitable matrix (a liquid solvent, a water/oil emulsion among others) at some concentration which can be sprayed on the skin or the clothes by the customers. After application of the product, the process of perception proceeds as follows: *i)* fragrances start to evaporate at different rates depending on their physicochemical properties and those of the mixture and, then, *ii)* they diffuse through the surrounding air until reaching the nostrils of consumers or other people around; and, finally, *iii)* fragrance molecules will eventually reach the human nose and above some concentration (known as the odor threshold) they will be detected and recognized with a specific intensity and character. Thus, our approach for Perfume Engineering combines knowledge from Chemical Engineering (Thermodynamics and Transport Phenomena) which can describe steps *i)* and *ii)*, together with models from Psychophysics for step *iii)* as shown in Figure 2.[6-9]

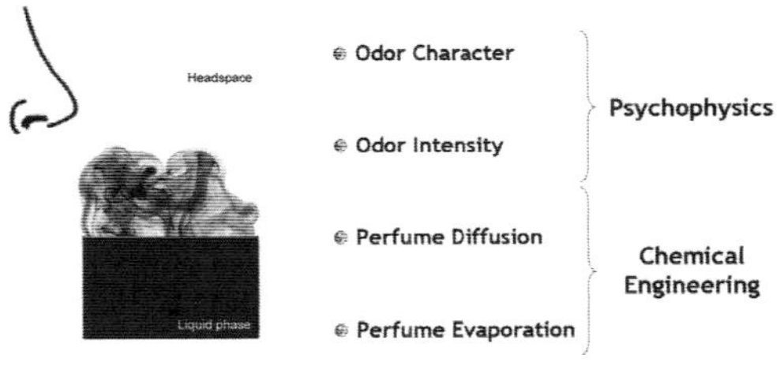

Figure 2 -- Schematic representation of our odor model for the perception of fragranced products.

Overall, our model will allow predicting and mapping the perceived odor elicited from any mixture of fragrance ingredients, thus reducing significantly the number of trial and error experiments which are expensive and time consuming for the Flavor & Fragrance industry.

In this work, we present a theoretical approach and experimental validation of our model for the step of fragrance diffusion together with the assessment of product performance using similar properties to assess it as the industry (like impact, tenacity, diffusion and volume parameters).[9] In this way, concentration profiles were experimentally measured in a diffusion tube, similar to the Stefan tube, and predicted with a model for pure fragrance chemicals, binary, quaternary, and multi-component (up to 11 chemicals) mixtures. A very good agreement between our purely predictive model and experimental concentration data was observed. Fragrance concentrations in air were then converted into odor intensities using models from psychophysics, making it possible to evaluate the evolution of the odor with time and distance using a performance plot. Accordingly, the performance of these mixtures was modeled and experimentally validated, which constitutes a landmark for fragrance design.

It was observed that the performance of a fragrance depends on the intrinsic properties of its constituents, but also on their molecular interactions in the liquid solution. Thus, the evaporation rate of each fragrance remains a function of mixture's initial composition (x_i), molecular interactions (activity coefficient, Ã¯Â □μ_i), and temperature (*T*) but it is simultaneously dependent on the physicochemical and psychological properties of each fragrance chemical (diffusivity, saturated vapor pressure, molecular weight, ODT, and power law exponent). Another important conclusion to draw from this work is that it is possible to measure and predict the performance of a perfumed product using perfumery terminology and a theoretical model to assess the perceived odor over time and distance from the source. Finally, it is expected that this work might contribute to reduce the time needed for the design of perfumes and fragranced products in the pre-formulation step and the reduction of the consumption of raw materials. Altogether, that will contribute to decrease production costs.

References

1. Cussler, E. C.; Moggridge, G. D., *Chemical Product Design*. Cambridge University Press: Cambridge, 2001.

2. Wesselingh, J. A.; Kiil, S.; Vigild, M. E., *Design & development of biological, chemical, food and pharmaceutical products*. Wiley: Chichester, U.K., 2007.

3. Cussler, E. L.; Wagner, A.; Marchal-Heussler, L., Designing Chemical Products Requires More Knowledge of Perception. *AIChE Journal* **2010,** 56, (2), 283-288.

4. Teixeira, M. A.; Rodriguez, O.; Gomes, P.; Mata, V.; Rodrigues, A., *Perfume Engineering: Design, Performance & Classification*. Elsevier: Oxford, UK, 2013.

5. Sell, C. S., On the unpredictability of odor. *Angewandte Chemie-International Edition* **2006,** 45, (38), 6254-6261.

6. Teixeira, M. A.; Rodriguez, O.; Mata, V. G.; Rodrigues, A. E., The Diffusion of Perfume Mixtures and Odor Performance. *Chemical Engineering Science* **2009,** 64, 2570-2589.

7. Teixeira, M. A.; Rodriguez, O.; Mata, V. G.; Rodrigues, A. E., Perfumery Quaternary Diagrams for Engineering Perfumes. *AIChE Journal* **2009,** 55, (8), 2171-2185.

8. Teixeira, M. A.; Rodriguez, O.; Mota, F. L.; Macedo, E. A.; Rodrigues, A. E., Evaluation of group-contribution methods to predict VLE and odor intensity of fragrances. *Industrial & Engineering Chemistry Research* **2011,** 50, 9390--9402.

9. Teixeira, M. A.; Rodriguez, O.; Rodrigues, A. E., Diffusion and Performance of Fragranced Products: Prediction and Validation. *AIChE Journal* **2013,** (accepted).

(269b) Qsrr Application to Two-Dimensional Gas Chromatography

Kamma, K., The University of Tokyo
Kaneko, H., The University of Tokyo
Funatsu, K., The University of Tokyo
Originally presented on: 11/5/2013 8:55:00 - 9:20:00

QSRR application to two-dimensional gas chromatography

1. Introduction

Gas chromatography is a widely used identification or elucidation method. In gas chromatography, a target sample is carried through a column filled with stationary phase, then retention time of compounds in the sample is observed. The structure of each compound in the sample can be identified based on its retention time, if the retention time of candidate compounds which are considered to be included in the sample is known from past observation data or from advance measurements. However, in this way, it is impossible to identify the structures of the compounds whose retention time is unknown.

Hence quantitative structure-retention relationship(QSRR)[1] is proposed to estimate the chromatographic retention time of any compounds. QSRR is a regression method of modeling the chemical structures and the retention time. First in structure descriptor calculation, chemical structures are converted into values which can be dealt with mathematically. Then a regression model is constructed between the retention time as an objective variable and the molecular structure descriptors as explanatory variables. In prediction, a new chemical structure is converted into structure descriptors and the constructed model predicts its retention time from the structure descriptors. A. Morsali et al. suggested QSRR models for BTEX and substituted benzenes based on chemical parameters.[2] Y. Zhang proposed QSRR models for polybrominated diphenyl ethers.[3]

Though QSRR is a useful method for the structure elucidation, large errors of the predicted retention time can be one of the crucial problems. Therefore it is important to construct a highly accurate model. But for the accurate models, prediction errors still remain to some degree. To deal with this problem, QSRR for two-dimensional chromatography[4] is considered useful. In two-dimensional chromatography, two different types of columns connected in line are used for retention time observation. The total retention time through the two columns is observed at the exit of second column, then retention time in each column is obtained. Constructing QSRR models for the retention time in each column enables the structure elucidation based on two predicted retention time. If one of the predicted retention time has large errors, the structure elucidation is still possible based on another predicted retention time.

The objective in this research is to improve the structure elucidation ability of two-dimensional gas chromatography by constructing QSRR models with high accuracy. We applied the partial least squares (PLS) method, the genetic algorithm-based PLS (GAPLS) method, ensemble PLS (EPLS) method to the data observed by two-dimensional gas chromatography. The prediction accuracy of the PLS, GAPLS and EPLS models is compared.

2. Method

2.1. PLS

PLS is a linear regression method. PLS extracts components which have strong correlation with objective variables from explanatory variables. Then the relationship between extracted components and objective variables is modeled. PLS can construct a robust model with high accuracy.

2.2. GAPLS

GA is an optimization method which imitates the evolution process. GAPLS is a variable selection method combining GA and PLS. GAPLS can efficiently optimize the combination of explanatory variables so that the prediction accuracy of the model improves.

2.3. EPLS

In EPLS, some variables are selected randomly from explanatory variables and many submodels are constructed between the selected variables and objective variables with the PLS method. Each submodel predicts values of objective variables, and mean or median of the predicted values are the final predicted values. The EPLS method can decrease the affection of prediction errors and increase the prediction accuracy of the model.

3. Result

3.1. Data

The set of retention time of 155 compounds was observed by two-dimensional gas chromatography. A non-polar column and a mid-polar one were used as the first column and the second column, respectively. The retention time in the first and second columns was set to be lower than 60 minutes and 8 seconds, respectively.

A software Dragon6[5] was used for structure descriptor calculation. The descriptors which failed to calculate were deleted. The descriptors in which more than 50% samples had the same value were deleted. Correlation coefficients between each descriptor are calculated, and one of the two descriptors whose correlation coefficient was over 0.9 was deleted. In the end, 164 descriptors were left.

100 compounds selected randomly were used as the training data, and the left 55 compounds were used as the test data.

3.2. QSRR Model Construction

The PLS, GAPLS and EPLS models were constructed with the training data and the models predicted the retention time of the test data. Table 1 shows the results. The retention time in the first and second column is represented as Y1 and Y2, respectively. R^2 is the evaluation criterion of the accuracy of the models. Q^2 is the cross-validated R^2 calculated by 5-fold cross-validation and R^2_{pred} is R^2 for the test data. When these values are close to 1, the model is accurate in prediction.

In prediction on Y1, R^2, Q^2 and R^2_{pred} values of PLS model were 0.984, 0.941 and 0.957, respectively. This result means that the prediction accuracy of PLS model for Y1 is high. The GAPLS model and the EPLS model also had high prediction accuracy for Y1.

In prediction on Y2, R^2, Q^2 and R^2_{pred} values of PLS model were 0.925, 0.743 and 0.676, respectively. These values mean that the prediction accuracy of PLS model is not very high. It is considered that some unnecessary variables in the explanatory variables decreased the prediction accuracy. For the GAPLS model, R^2, Q^2 and R^2_{pred} values were 0.929, 0.882 and 0.754, respectively. R^2_{pred} value of the EPLS model was 0.740, which was close to that of the GAPLS model. It can be said that the GAPLS and EPLS methods obviously improved the prediction accuracy.

In any models, the prediction accuracy for Y2 was lower than that for Y1. As a reason for that, it is considered that the observation errors of Y2 decreased the prediction accuracy. Because the second column was short so that the measurements took less than 8 seconds, observation errors of Y2 could be large.

Table 1. Modeling and prediction result

	Y1			Y2		
	R^2	Q^2	R^2_{pred}	R^2	Q^2	R^2_{pred}
PLS	0.984	0.941	0.957	0.925	0.743	0.676
GAPLS	0.990	0.983	0.966	0.929	0.882	0.754
EPLS	-	-	0.953	-	-	0.740

4. Conclusion

We aimed to construct highly accurate QSRR models for two-dimensional gas chromatography by using the PLS, GAPLS and EPLS methods. As a result, we could construct the Y1 and Y2 prediction models having high performance although the prediction accuracy of Y2 was lower than that of Y1.

In this research, we assumed that the candidate compounds included in the target sample are already presumed. But mostly, it is difficult to presume the candidate compounds in advance. In the case that candidate compounds are totally unknown, inverse analysis of QSRR models can be applied to predict candidate chemical structure based on retention time and the constructed QSRR models. It is expected that inverse analysis achieves more useful structure elucidation.

5. Acknowledgement

The authors acknowledge the support of Asahi Kasei Corp. for offering the data of two-dimensional gas chromatography.

6. References

[1] R. Kaliszan. Structure and Retention in Chromatography A Chemometrics Approach. CRC Press. May 22, 1997

[2] A. Morsali, S. A. Beyramabadi, M. R. Bozorgmehr, M. Raanaee, B. Keyvani and G. R. Jafari. Prediction of gas chromatography retention of BTEX and other substituted benzenes based on quantum chemical parameters. Scientific Research and Essays Vol. 5 (3), pp. 349-351, 4 February 2010

[3] Ya-Hui Zhang, Shu-Shen Liu, Hong-Yan Liu. Predicting the Gas Chromatographic Relative Retention Time of Polybrominated Diphenyl Ethers by MEDV-13 Descriptors. Chromatographia Vol. 65, Issue 5-6, pp 319-324, March 2007

[4] T. Hyötyläinen. Comprehensive Two-Dimensional Chromatography. SPRINGER, BERLIN. February 2013

[5] Talete srl. http://www.talete.mi.it/ 13 May 2013

(269c) Method of Moments for Computational Microemulsion Analysis and Prediction in Tertiary Oil Recovery

Fraaije, J. G. E. M., Culgi BV
Tandon, K.,
Jain, S.,
Handgraaf, J. W., University of Trieste
Buijse, M.,
Originally presented on: 11/5/2013 9:20:00 - 9:45:00

We discuss the application of Helfrich's surface torque density concept to microemulsion design and analysis, from three different angles: (i) from the point of view of coarse-grained molecular simulations, using Dissipative Particle Dynamics, including charge interactions and added salt, (ii) using an approximate double film model for the surface, and (iii) comparison with formulation approaches. The simulations use that surface torque can be calculated unambiguously from the stress profile, provided the surface is tensionless. Very good agreement is found on predicting optimal salinity (or the absence of that) for a range of surfactants: Dioctyl Sodium Sulfosuccinate (AOT), various twin tailed sulfonates and Sodium Dodecyl Sulfate (SDS). The simulations are very fast, on par with times for experiments, thus they could lead to a practical tool for discovery of more efficient surfactants, although much remains to be done with respect to other important variables: oil composition, surfactant mixtures, aggregation in solution and so on. The microscopic model (second approach) is highly approximate: it is essentially based on two opposing swelling tendencies, that are both of osmotic nature. According to the model, the tails are swollen by the oil, and the charged head groups are confined in a salty layer in Donnan equilibrium with the salt solution. In this way, the surface interactions are purely entropic. The comparison of the film model with existing formulation approaches (third approach) covers: interfacial tension minimum, Winsor R theory, Quantitative Structure Property Relations (QSPR), Hydrophilic-Lipophilic Deviation (HLD), and HLD-Net Average Curvature, and temperature coefficients. Using the surface torque analysis, we succeed in deriving in an 'ab initio' way QSPR empirical coefficients that have been known for decades, but, until now, have been obscure in origin.

(269d) A Computational Methodology for Designing Mixtures

Austin, N.,
Samudra, A., Rockwell Automation
Sahinidis, N., Carnegie Mellon University
Trahan, D. W., Massachusetts Institute of Technology
Originally presented on: 11/5/2013 9:45:00 - 10:10:00

As trial-and-error methods are becoming increasingly intractable with the growing library of chemical compounds, computer-aided molecular design (CAMD) is a necessary tool for the future of chemical production. CAMD combines optimization techniques with many property prediction models, allowing for a rapid and accurate search through all chemical structures, even those for which no data is available. These techniques work well for designing single molecules, but a more difficult –and more applicable – problem arises when attempting to design mixtures [1].

We propose a novel mixture design approach which couples molecular design with derivative-free and algebraic model-based optimization. We use derivative-free optimization (DFO) to search in the space of compound properties. Requirements on properties of the individual components are then converted into molecular structures via our software AMODEO and a recently developed molecular design methodology [2], which relies on Group Contribution methods for property estimation [3]. Finally, we use mixing rules and algebraic global optimization in order to optimize over the composition space for a given set of mixture components. The approach exploits a key insight to the molecular mixture design problem, namely that the small number of degrees of freedom in the property space can be dealt with efficiently via derivative-free optimization (DFO) algorithms [4]. Several case studies are presented to illustrate the wide applicability of the proposed methodology.

References:

1. Conte, E., R. Gani, and K. M. Ng, Design of formulated products: A systematic methodology, AIChE Journal, 57, 2431-2449, 2011
2. Samudra, A. and N. V. Sahinidis, Optimization-based framework for computer-aided molecular design, AIChE Journal, accepted, 2013
3. Marrero, J. and R. Gani. Group-contribution based estimation of pure component properties, Fluid Phase Equilibria, 183–184:183–208, 2001
4. Rios, L. M. and N. V. Sahinidis, Derivative-free optimization: A review of algorithms and comparison of software implementations, Journal of Global Optimization, DOI 10.1007/s10898-012-9951-y, 2012

(269e) Developing a QSAR Relating Solvent Structure to Crystal Morphology Using 2D & 3D Molecular Descriptors

Haser, J. C., Auburn University
Herring, R. III, Auburn University
Eden, M. R., Auburn University
Originally presented on: 11/5/2013 10:10:00 - 10:35:00

Crystallization is one of the least-studied separation techniques compared to distillation and extraction processes, especially in computer modeling and prediction. There are many more variables in determining final quality (crystal morphology, crystal purity, potential recovery) in crystallization processes than distillation and extraction [1]. Crystallization process products are typically produced in small quantities and their properties can have a large effect on the downstream processes. Therefore it is important that crystals meet strict guidelines in final quality.

Studies have been done to predict ibuprofen crystal morphologies using linear models based on hydrogen bonding properties of several different pure solvents [2]. The goal of this work is to develop a predictive model of crystal morphology based on solvent choice. The model will be developed using Principal Component Analysis and multivariable linear regression. Using a combination of 2D & 3D molecular descriptors is ideal for crystallization solvent design because the structure of the solvent is critical to the resulting morphology of the crystal [2]. With this model, the relationship between solvent molecular geometry and crystal aspect ratio can be determined. The first work will be focused on the aspect ratio of ibuprofen crystals and then further expanded to account for any solute molecule. It is advantageous to be able to determine what solvent characterization will produce crystals of the desired quality, and this method will allow solvents to be chosen without using time- and money-consuming experimental measurements.

[1] Karunanithi A.T., Achenie, L.E.K, and Gani, R. (2006) A computer-aided molecular design framework for crystallization solvent design. Chem. Eng. Sci., **61**, pp. 1247–1260.

[2] Acquah, C., Karunanithi, A.T., Cagnetta, M., Achenie, L.E.K., and Suib, S.L. (2009) Linear models for prediction of ibuprofen crystal morphology based on hydrogen bonding propensities. Fluid Phase Equilibria, **277**, pp. 73-80.

(269f) Achieving Target Emulsion Drop Size Distributions Using Population Balance Equation Models of High Pressure Homogenization

Maindarkar, S. N., University of Massachusetts
Dubbelboer, A., University of Technology
Hoogland, H., Unilever R & D
Henson, M. A., University of Massachusetts Amherst
Originally presented on: 11/5/2013 10:35:00 - 11:00:00

Oil-in-water emulsions are ubiquitous dispersed phase systems with diverse applications that include consumer products, processed foods, polishes, waxes, agricultural sprays and road surfacing materials. In the foods industry, emulsions constitute natural foods as well as numerous processed products such as milk, butter, margarine, ice cream, sauces and desserts. Food emulsions contain edible oils, water and biocompatible surfactants as the major ingredients and vitamins, minerals and/or flavors as minor ingredients. Emulsion system formulation and processing operations both impact the drop size distribution, a key property that influences emulsion rheology, stability, texture and appearance. A typical processed food requires the drop size distribution to be maintained within acceptable limits, which includes achieving a prescribed mean drop size, maintaining small variations about the mean and avoiding very small or large drops that adversely affect product properties such as texture, appearance and food safety. Oil-in-water emulsions are typically formed by first preparing a coarse premix using a low shear stator-rotor type device that mixes the various ingredients into a stable form. This premix is then processed with a high shear device in which relatively large drops are broken into much smaller drops. In high pressure homogenization, the coarse emulsion is passed through a small orifice under very high pressure.

Due to lack of quantitative understanding, new emulsified products are currently developed by combining a broad knowledge of previous product formulations with empirical scientific experimentation. An alternative to brute force experimentation is to utilize a suitable mathematical model to predict the drop size distribution for different emulsion formulations and processing conditions. The population balance equation (PBE) modeling framework is particularly well suited for this problem as functions describing drop breakage and coalescence can be incorporated within a fundamental number balance equation to predict the evolution of the drop size distribution. Several investigators have developed PBE models of high pressure homogenizers. As part of our previous work, we have developed breakage-only PBE models for prediction of the drop volume distribution. Through experimental and computational studies, we showed that our model emulsion system exhibited negligible coalescence due to the low oil-to-surfactant ratio (5% oil, 1% surfactant) used.

In this paper, we incorporate drop coalescence functions to allow the drop volume distribution to be predicted for a much higher oil-to-surfactant ratio (50% oil, 1% surfactant) and utilize the model to achieve target distributions by changing homogenizer operating conditions. Nonlinear optimization is used to estimate six adjustable parameters in the drop breakage and coalescence functions from measured size distributions. The experimentally validated model is used to optimally determine the number of homogenization passes and the pressure of each pass such that specified emulsion drop size properties are achieved in a least-squares sense. Two optimization objectives that differ with respect to the distribution specifications are formulated and experimentally evaluated.

(272a) On the Location of Phase Boundaries in Alloys Using Multi-Phase and Multi-Cell Simulations

Gelb, L. D., University of Texas at Dallas
Wang, Y.,
Originally presented on: 11/5/2013 8:30:00 - 8:48:00

Embedded-atom models (EAMs) are empirical many-body potentials widely used in studies of bulk metals, alloys, films, and nanoparticles. EAMs are largely parameterized again solid-state properties, with occasional inclusion of small-cluster data and/or melting points. EAM melting points are commonly obtained using two-phase molecular-dynamics simulations at constant total energy, though with simulation protocols of varying accuracy. We investigate the various factors influencing the results of two-phase simulations and show that, with proper treatment of stress, results accurate to four significant figures or better can be reliably obtained. Finite size dependence in selected models is also analyzed and found to be nontrivial, in contrast with some previous work; nonetheless, extrapolation to the thermodynamic limit can successfully address this problem. We also demonstrate reliable methods for the location of solid-solid-liquid (eutectic) and other triple points in alloys systems, which are of particular interest in practical applications. These are based on extensions of the Gibbs ensemble method, consisting of Monte Carlo and hybrid MC/MD simulations of multiple coupled simulation cells.

(272b) Successive Global and Local Optimization Techniques Utilized to Generate New Accurate VLE Force Fields

Kramer, A.,
Hulsmann, M., Fraunhofer Institute for Algorithms and Scientific Computing (SCAI)
Reith, D., Bonn-Rhein-Sieg University of Applied Studies
Originally presented on: 11/5/2013 8:48:00 - 9:06:00

The practical applicability of atomistic soft matter simulations requires the construction of appropriate molecular models for a wide range of chemicals. The key to a quantitative property prediction is the accuracy of the simulation's basis, i.e. the force field. This is particularly true for the intermolecular interactions. Manual adjustment and optimization of the respective parameters is, at best, extremely time consuming. Hence, semi-automated parameterization schemes are essential in the pursuit to create tailor-made models for specific molecules in a timely fashion.

In this contribution, we want to introduce and assess global optimization techniques as new crucial components for the intermolecular part of our force field calibration workflows. Two new Vapor--Liquid--Equilibrium (VLE) force fields for ethylene oxide are presented as a test case. Both of them were developed by a combination of highly efficient global and local optimization algorithms. Thereby, a weighted quadratic loss function between simulated and experimental VLE properties at different temperatures, i.e. saturated liquid density, enthalpy of vaporization, and vapor pressure, is minimized. The global optimization techniques considered here are based on the collocation of meta models of either the loss function itself or of each of the VLE properties. We will present, how the application of existing methods led us to our own algorithm, called COSMoS („Calibration/Optimization by Simultaneous Modeling of Simulated data"), and how it performs.

The resulting force fields reproduce the saturated liquid density with a maximum error of 0.5%, the enthalpy of vaporization with a maximum error of 1%, and the vapor pressure with a maximum error of 5%. Furthermore, the models are capable to predict other thermo-physical properties quite well.

(272c) Recent Developments of Highly-Scalable Coarse-Grain Modeling Tools Formulated Upon Dissipative Particle Dynamics

Larentzos, J. P., Sandia National Laboratories
Moore, J. D., U.S. Army Research Laboratory
Lisal, M., Academy of Sciences of the Czech Republic
Mattson, W.,
Brennan, J. K., U.S. Army Research Laboratory
Originally presented on: 11/5/2013 9:06:00 - 9:24:00

Modeling and simulation of materials at the micro- and mesoscales have a wide range of challenges. For example, the mechanical properties of many polymers are dominated by topological constraints or entanglements, which prohibit the use of atomistic-level modeling and simulation. Therefore, the simulation of the morphology and mechanics of microphase separated systems, such as gels and polymer networks require mesoscale modeling and simulation. One such method, Dissipative Particle Dynamics (DPD), has become a common method of choice for use in the study of coarsened soft matter (e.g., surfactants, polymers, biomolecules). While DPD itself is isothermal, the DPD concepts and methodology have been extended to simulate condensed-phase systems under isothermal, isobaric, isoenergetic, and isoenthalpic conditions.

Numerical integration of the DPD equations-of-motion requires special consideration, where the Shardlow-splitting algorithm (SSA) has been found to be the most appropriate integration scheme available to-date. Despite significant DPD algorithmic improvements with the SSA, there are considerable challenges in parallelizing the SSA algorithm. This work focuses on exploiting multi-core architectures by converting the serial SSA algorithm into a highly-scalable application-focused software package. The multi-scale suite of tools has been implemented into the open-source LAMMPS (Large-scale Atomic/Molecular Massively Parallel Simulator) simulation package. While the suite of DPD methods has major application in polymer and biological sciences, we will demonstrate that they also can be used for molecular solids. In particular, we will demonstrate shock and heating of coarsened RDX (cyclotrimethylenetrinitramine), an energetic material, where dynamic and thermodynamic properties using the constant-energy DPD method agree well with fully atomistic simulations.

(272d) A Recent Advance in Modeling Dynamic Properties With Coarse-Grained Molecular Simulation

Markutsya, S., Iowa State University
Lamm, M. H., Iowa State University
Originally presented on: 11/5/2013 9:24:00 - 9:42:00

Coarse-grained models that are rigorously derived from all-atom molecular simulation trajectories can be successfully implemented in molecular dynamics simulation to reproduce the structure and thermodynamics of the all-atom reference system. Dynamic properties from these coarse-grained molecular simulations are often unreliable because the reduction in degrees of freedom eliminates much of the friction between coarse-grained beads. This results in faster dynamics for the coarse-grained simulation. This is an advantage when it comes to efficient sampling of phase space but problematic if accurate computation of one or more time-dependent properties is a desired outcome from the coarse-grained simulation. In this work, a new approach is developed for deriving coarse-grained intermolecular forces. This approach retains the frictional contribution that is often discarded by conventional coarse-graining methods. This method is based on the well-known Langevin equation formalism. Compared to previous implementations of the Langevin equation for coarse-grained dynamics, in this method, the friction coefficients are computed directly during the derivation of the CG force field, using routine post-processing calculations. The procedure is tested for water and an aqueous glucose solution, and the results from the new implementation for coarse-grained molecular dynamics simulation show remarkable agreement with the dynamics obtained from reference all-atom simulations. The agreement between the structural properties observed in the coarse-grained and all-atom simulations is also preserved. We discuss how this approach may be applied broadly to any existing coarse-graining method where the coarse-grained models are rigorously derived from all-atom reference systems.

This research is sponsored by the U.S Department of Energy (USDOE) Scientific Discovery through Advanced Computing (SciDAC) program through USDOE's Office of Advanced Scientific Computing Research (ASCR) and Biological and Environmental Research (BER), and performed at the Ames Laboratory, FWP AL-08-330-039.

(272e) A Systematic Coarse-Graining of Molecular Dynamics Simulations: Thermodynamic and Transport Properties

Fu, C. C., University of California, Santa Barbara
Kulkarni, P. M., The City College of New York
Shell, M. S., University of California, Santa Barbara
Leal, L. G., University of California, Santa Barbara
Originally presented on: 11/5/2013 9:42:00 - 10:00:00

Coarse-graining techniques have recently attracted great interest for providing descriptions at a mesoscopic level of resolution that preserve fluid thermodynamic and transport behaviors with a reduced number of degrees of freedom (DOFs) and hence less computational effort. One fundamental question arises: how well and to what extent can a "bottom-up" developed mesoscale model recover the physical properties of a molecular scale system? To answer this question, we explore *both* the thermodynamic and transport properties of a systematically-developed coarse-grained (CG) model that represents an intermediate mesoscale fluid between the atomistic and continuum scales [1,2]. The model is developed using the iterative Boltzmann inversion (IBI) technique to determine a CG potential for a $(1-\varphi)N$ mesoscale particle system, where φ is the fraction of removed particles from an atomic system. The uniqueness theorem guarantees a one to one relationship between the radial distribution function (RDF) and such effective pairwise potentials, but we find that RDFs are insensitive to the long-range part of the IBI-determined potentials, which provides significant flexibility in further matching other properties.

We subsequently propose a reformulation of IBI that enables simultaneous matching of the RDF and the fluid pressure. This new method mainly changes the attractive tail region of the CG potentials, and it improves the isothermal compressibility relative to pure IBI. We also find that there are optimal interaction cutoff lengths for the CG system, as a function of φ, that are required to attain an adequate potential while maintaining computational speedup. Dynamical properties such as the self diffusion coefficient and viscosity cannot be matched directly during coarse-graining by modifying the pair interaction. Instead, one can introduce a dissipative and random forces characterized by a friction coefficient γ, which becomes an additional parameter in the CG model that can be tuned. Using the Galilean-invariant dissipative particle dynamics thermostat, we show that a value of γ for each degree of coarse-graining φ can be found for which both viscosity and diffusion match the reference LJ liquid. Importantly, we show that Stokes-Einstein behavior persists for these coarse models and offers a useful perspective for interpreting the dynamics of mesoscale CG models.

References:

[1] Fu, C.; Kulkarni, P.; Shell, M. S.; Leal, L. G. *J. Chem. Phys.* **137**, 164106 (2012).

[2] Fu, C.; Kulkarni, P.; Shell, M. S.; Leal, L. G. *J. Chem. Phys.* 2013, Submitted.

(272f) Coarse-Grained Lattice Monte Carlo Simulations With Continuous Interaction Potentials

Sinno, T., University of Pennsylvania
Liu, X., University of Pennsylvania
Seider, W. D., University of Pennsylvania
Originally presented on: 11/5/2013 10:00:00 - 10:18:00

Various types of Monte Carlo simulations are used to simulate an enormous range of phenomena. Restricting particle positions to fixed lattice sites can substantially increase the computational efficiency of Monte Carlo simulations, and this benefit increases as the lattice becomes coarser. However, the confinement of particle positions to a rigid lattice necessarily reduces the available configurational degrees of freedom in a system and this constraint can become very important at elevated temperatures [1,2]. As a result, any coarse-graining approach that maps a continuous problem onto a lattice must be able to implicitly describe the influence of the missing degrees-of-freedom.

In this presentation, we discuss a new framework for performing Metropolis Monte Carlo simulations of continuous systems on coarse, rigid lattices while preserving the phase-space contributions of the missing degrees-of-freedom [3,4]. The present approach relies on the pre-computation of (approximate) discrete coarse-grained interaction potentials from standard molecular force-fields. The coarse-grained potentials are then used to drive coarse-grained Metropolis Monte Carlo (CG-MMC) simulations of vapor-liquid systems based on Lennard-Jones, square-well, and SPC-water models. The method accuracy, scalability, and efficiency are first demonstrated via consideration of vapor-liquid equilibria. We then extend the CG-MMC approach to non-equilibrium situations in which diffusion and advection are both present.

[1] S. S. Kapur, M. Prasad, J. C. Crocker, T. Sinno, Role of configurational entropy in the thermodynamics of clusters of point defects in crystalline solids, Phys. Rev. B **72**, 014119 (2005).

[2] J. Dai, W. D. Seider and T. Sinno, Lattice kinetic Monte Carlo simulations of defect evolution in crystals at elevated temperature. Molecular Simulation **32**, 305 (2006).

[3] X. Liu, W. Seider, and T. Sinno, Coarse-Grained Lattice Monte Carlo Simulations with Continuous Interaction Potentials, Physical Review E **86**(2012) 026708 1-5.

[4] X. Liu, W. Seider, and T. Sinno, A General Method for Spatially Coarse-Graining Metropolis Monte Carlo Simulations onto a Lattice, Journal of Chemical Physics **138** (2012) 114104 1-14.

(272g) Multicanonical Simulations Based On Contact Maps for Square-Well Heteropolymers Using Event Driven Molecular Dynamics

McGuire, A., University of Strathclyde
Bannerman, M. N., University of Aberdeen
Curtis, R. A., The University of Manchester
Lue, L., University of Strathclyde
Originally presented on: 11/5/2013 10:18:00 - 10:36:00

The folding of proteins into a unique, specific 3-dimensional structure is driven primarily by the hydrophobic effect, with hydrophobic groups buried in the interior of the molecule, away from contact with water, and hydrophilic groups on the surface of the molecule, solvated with water. It is the ability of the protein to fold into its native conformation that allows it to function (e.g., catalyze a biochemical reaction). While the hydrophobic effect drives protein folding, it also can lead to protein misfolding (trapped in a non-native conformation) and, in protein solutions, to aggregation, where the hydrophobic groups from distinct proteins adhere to each other. The misfolding and aggregation of protein solutions is thought to be the basis of Alzheimer's and Parkinson's diseases. They also lead to difficulties for bioprocessing, such as in the recovery and purification of biopharmaceuticals.

The delicate competition between intramolecular and intermolecular interactions determines whether the proteins in a solution will remain folded or aggregate. While simulations of detailed, fully atomistic models have significantly advanced our understanding of protein behavior, they are too computationally intensive to directly apply to the aggregation problem. Coarse-grained protein models, where the protein is represented by a collection of bonded hard spheres that interact with each other with a square-well potential, are simple enough to offer the possibility of simulating bulk protein solutions but are sufficiently complex to capture the essential physics of protein behavior (e.g., folding) [1]. By varying the monomer size, bond length, and range of interaction, these off-lattice models can exhibit intricate three-dimensional structures, such as helices [2,3] and other more complicated configurations.

The use of a discontinuous interaction potential (i.e., hard sphere plus a square-well attraction) greatly simplifies the integration of the equation of motion. We have recently developed an extremely efficient event driven molecular dynamics simulation package, DynamO [4], which allows the study of large systems and sufficiently long time scales needed to describe the different stages of protein folding and aggregation on relatively modest computing hardware. These large-scale simulations are necessary, in order to properly investigate the coupling between the intra and intermolecular interactions.

In order to determine the equilibrium behavior of these systems, it is crucial that the important configurations of the system can be rapidly and properly sampled. Even these relatively simple coarse-grained models possess a fairly rugged free energy landscape, and conventional canonical Monte Carlo and molecular dynamics simulations have limited
applicability. Simulation methods have been developed to try to overcome the difficulties associated with rough free energy landscapes, such as the replica exchange method and multicanonical simulations.

For monomers which interact through square-well potentials, the structure of the molecule can characterized by its contact map, a list of monomer pairs that are located within each other's interaction well. In this work, we develop a multicanonical simulation technique where distinct contact maps are sampled with equal likelihood by associating different multicanonical weights to each contact map. This has the advantage over conventional multicanonical simulations, where different energies are sampled with equal probability, because the use of contact maps gives the ability to distinguish between different folded states with the same energy and allows us to label and follow the folding pathway of the molecules. In addition, it allows the determination of the free energy associated with a particular contact map. These multicanonical contact map simulations were performed using event driven molecular dynamics by altering the collision rules between monomers. This has been implemented within DYNAMO. In addition, an algorithm has been develop to iteratively determine value of the multicanonical weights.

To test this method, we first determine the density of states of homopolymer models, which exhibit a coil-helix transition, and compare
the efficiency with other simulation techniques. Then we examine the structure and folding of a series of protein sequences, focusing on characterizing the folded state of the proteins, with particular attention paid to its entropy. Finally, the interaction between pairs of protein molecules are examined to understand the influence of intermolecular interactions on the folding of the individual proteins.

[1] JE Magee, J Warwicker, and L Lue, "Freezing and folding behavior in simple off-lattice heteropolymers," J. Chem. Phys. 120, 11285

(2004).
[2] JE Magee, VR Vasquez, and L Lue, "Helical structures from an isotropic homopolymer model," Phys. Rev. Lett. 96, 207802 (2006).
[3] MN Bannerman, JE Magee, and L Lue, "Structure and stability of helices in square-well homopolymers," Phys. Rev. E 80, 021801 (2009).
[4] MN Bannerman, R Sargant, and L Lue, "DynamO: A free O(N) general
event-driven molecular dynamics simulator," J. Comput. Chem. 32, 3329 (2011).

(272h) Accelerating Screening of Simulation Parameter Spaces By Orders of Magnitude Using Reweighting and Configuration Space Mapping Algorithms

Paliwal, H., University of Virginia
Shirts, M. R., University of Virginia
Originally presented on: 11/5/2013 10:36:00 - 10:54:00

Classical force field parameterization is computationally intensive. The fitting procedure involves computationally expensive molecular simulations to estimate observables at each iteration. The new set of parameters is either guessed based on a trial and error approach or is generated by symplectic or gradient based optimization. The time spent in either case is dominated by evaluation of the objective function which requires fresh generation of samples for a new guess of parameter set. We can accelerate the force field parameterization process if we can reduce the time spent in sampling at each iteration. In this work we present a parameterization scheme. We gain speed by skipping several steps of molecular simulation by replacing these runs using multistate reweighting techniques to estimate the observables. The multistate reweighting formalism requires re-evaluation of the energies with the new set of parameters using the configurations generated in simulations run with the initial guess set of parameters. The time consumption for estimating observables using re-evaluated energies is orders of magnitude less compared to a fresh simulation. By incorporating configuration space transformations, we can reweight over changes in geometric parameters, even for rigid molecular models, as well as energetic parameters. We demonstrate the utility and efficiency of this parameterization scheme by optimizing over the space of noncovalent simulation parameters in free energy calculations, as well as exploring the parameter space of liquid water models.

(329a) Contact-Value Self-Consistency in Density Functional Theory Calculations

Jiang, J.,
Cao, D., Beijing University of Chemical Technology
Wu, J., UC Riverside
Originally presented on: 11/5/2013 12:30:00 - 12:48:00

Density functional theory (DFT) calculations are typically based on an approximate functional that connects the properties of a multi-body system with the underlying one-body density distributions. The state of art for classical systems is to represent the functional in terms of the fundamental measure theory for the molecular excluded volume effects and the Gaussian perturbation for various components of the long-range intermolecular forces. Unlike the conventional local-density or mean-field approximations, the more sophisticated functional is able to reproduce simulation results quantitatively without the significant increase of the computational burden. Nevertheless, any approximation introduced in the functional formulation may violate the statistical-mechanical sum rules leading to thermodynamic inconsistency. In this work, we introduced a new computational procedure for the implementation of the DFT calculations that conform exact statistical-mechanical relations. The numerical performance of the new theoretical method has been demonstrated by its application to predicting the structure of electric double layers within the standard electrolyte models. Drastic improvement has been found in comparison of the DFT calculations with the simulation results.

(329b) From Virtual Catalysis to Practical Applications: Understanding and Designing Catalysts in Silico

Stamatakis, M., University of Delaware
Originally presented on: 11/5/2013 12:48:00 - 13:06:00

The importance of heterogeneous catalysis in modern applications that enhance the quality of life cannot be overstated: current estimates place the value of the catalyst market at more than $20 billion, and it has been estimated that every $1 spent on a catalyst can generate up to $1000 worth of product. Computational methods are increasingly becoming crucial in gaining a fundamental understanding of the physicochemical phenomena that give rise to catalytic activity. Such an understanding makes it possible to design improved catalysts; however, the complexity encountered in catalytic systems poses several challenges in modelling them.

We present an *ab-initio* kinetic Monte Carlo (KMC) framework that uses concepts from graph theory to overcome these challenges and easily treat complex chemistries and surface phenomena. This approach allows one to define elementary reaction events characterized by complex neighbouring patterns of the sites involved. These sites can be of various types (such as steps or terraces) and occupied by monodentate as well as multidentate species. It also treats adsorbate-adsorbate lateral interactions in a generalised way by incorporating cluster expansions for the energetics.

We further showcase this approach by applying it to key catalytic reactions, in particular CO oxidation on Au nano-clusters, which is important for environmental and health reasons, and water-gas shift on single crystal Pt surfaces, which is relevant to H_2 production from biomass. We show that our approach can provide a wealth of information, which can be used to explain experimental observations and obtain insight in the design of materials with desired catalytic properties.

(329d) From Structure to Thermodynamics With Local Molecular Field Theory

Remsing, R. C., University of Maryland
Weeks, J. D., University of Maryland
Originally presented on: 11/5/2013 13:24:00 - 13:42:00

A fundamental problem in statistical mechanics is the development of accurate and efficient means to calculate free energies. A novel "bottom-up" approach to the calculation of thermodynamic properties is described, in which free energies of systems containing long-ranged interactions are obtained by integration over the structure of an appropriately chosen purely short-ranged "mimic system", a fact that is especially advantageous when long-ranged Coulomb interactions are present. Local molecular field (LMF) theory of nonuniform fluids is employed to obtain an accurate description of the structure in the short ranged system, then LMF ideas are extended in a very simple way to determine contributions to the free energy from the long ranged interactions using only the structural properties of the mimic system. After introducing this LMF theory-based framework, its accuracy will be demonstrated on several problems of interest, including hydration free energies of hydrophobic solutes and ions, simple alchemical transformations, and general potentials of mean force.

(329e) Multilevel Summation for Dispersion: A Linear-Time Algorithm for 1/R6 Potentials

Tameling, D.,
Springer, P., RWTH Aachen University
Bientinesi, P., RWTH Aachen University
Ismail, A. E., RWTH Aachen University
Originally presented on: 11/5/2013 13:42:00 - 14:00:00

The multilevel summation method (MLS) was developed to evaluate of long-range interactions in molecular dynamics (MD) simulations. Previously MLS was investigated in detail for the electrostatic potential by Hardy et al., and we have applied this new method to dispersion interactions. While dispersion interactions are formally short-ranged, long-range methods have to be used to calculate accurately dispersion forces in certain situations, such as in interfacial systems. Because long-range solvers tend to dominate the time needed to perform a step in MD calculations, increasing their performance is of vital importance.

The multilevel summation method offers some significant advantages when compared to mesh-based Ewald methods like the particle-particle particle-mesh and particle mesh Ewald methods. Because, unlike mesh-based Ewald methods, the multilevel summation method does not use fast Fourier transforms, they are not limited by communication and bandwidth concerns. In addition, it scales linearly in the number of particles, as compared to the $O(N \log N)$ complexity of the mesh-based methods.

While the structure of the Multilevel Summation is invariant for different potentials, every algorithmic step had to be adapted to accommodate the r^{-6} form of the dispersion interactions. In addition, we have derived strict error bounds, similar to those obtained by Hardy for the electrostatic multilevel summation method. Using an unoptimized implementation in C++, we can already demonstrate the linear scaling of the multilevel summation method for dispersion, and present results that suggest it will be competitive in terms of accuracy and performance with mesh-based methods.

(329f) A Linear Basis Function Approach to Efficient Molecular Transformation Free Energy Calculations

Naden, L. N., University of Virginia
Shirts, M. R., University of Virginia
Originally presented on: 11/5/2013 14:00:00 - 14:18:00

We present a formalism for inserting or removing atomic sites in free energy calculations based on splitting the potential energy function into a sum of pairs of basis functions, which depend only on coordinates, and alchemical switches, which depend only on the coupling variable. This formalism allows evaluation of nonbonded potentials at any intermediate state accessible to the potential outside of the simulation's force evaluation loop or entirely in post-processing, which may be particularly useful to minimize effort in coding for in complex architectures such as GPU's.

This method can design optimal minimal variance switches to reduce statistical uncertainty using a single simulation's worth of data for a given transformation. We show examples of this approach by minimizing the statistical uncertainty for the solvation free energy of a wide range of molecule types and sizes, from united atom spheres with sizes ranging from methane to buckmisterfullerenes, to large multi-ring aromatics and ions, as well as optimal transformations of one molecule type to another.

We find that the statistical error in free energy for the optimized basis function approach are lower than standard soft core models, and approach the minimum variance possible over all pair potentials. The basis function approach can easily be applied to essentially all systems, reduce code complexity, and reduce simulation times.

(329g) Study of Phase Transitions By 2-D Density of States Simulation

Gai, L., Vanderbilt University
Iacovella, C. R., Vanderbilt University
Cummings, P. T., Oak Ridge National Laboratory
McCabe, C., Vanderbilt University
Originally presented on: 11/5/2013 14:18:00 - 14:36:00

We have developed a hybrid Wang-Landau (WL)/statistical temperature molecular dynamics (STMD) method to simulate a 2D joint density of states (JDOS). The 2D JDOS [1] of continuous models is an important quantity as the free energy can be calculated not only as a function of temperature, but also as a function of concentration or pressure. However, determining the JDOS requires significant simulation time and is impractical with the original Wang-Landau (WL) method [2]. Previously, we have examined the phase behavior of self-assembling lipids as a function of temperature, *via* the statistical temperature molecular dynamics (STMD) [3] and demonstrated improved efficiency compared to the original WL method. STMD is an extension of the WL method that is suitable for simulations with large energy bins and can generate a deterministic MD trajectory to extend the WL method to complicated systems beyond the scope of effective MC moves. Building on earlier work [4-5], we have developed a hybrid WL /STMD method to simulate the 2D JDOS of energy and lipid concentration in the semi-grand canonical ensemble. The lipid model analyzed consists of 3 coarse-grained segments solvated by water. The phase behavior of the lipids with respect to bilayer formation has been characterized through the calculation of the heat capacity as a function of both temperature and lipid concentration. The method enables simulations to be performed to very low temperatures, with the whole system exhibiting well-ordered structures. A full view of the self-assembling bilayer phases are observed within the temperature range studied along each concentration, including gel phase bilayers with frozen water, with mixed water (i.e., frozen and liquid water), and with liquid water and a more fluid bilayer with liquid water. The results obtained at selected concentration are compared to those from STMD with independent runs at the same concentration, and are consistently in excellent agreement with each other.

1. C. Zhou, T.C. Schulthess, S. Torbrügge, and D.P. Landau, "Wang-Landau algorithm for continuous models and joint density of states," *Phys. Rev. Lett.* **96** (2006) 120201

2. F. Wang and D. P. Landau, "Efficient, Multiple-Range Random Walk Algorithm to Calculate the Density of States," *Phys. Rev. Lett.* **86** (2001) 2050–2053.

3. J. Kim, J. E. Straub and T. Keyes, "Statistical-Temperature Monte Carlo and Molecular Dynamics Algorithms," *Phys. Rev. Lett.* **97** (2006) 050601

4. L. Gai, K. Maerzke, P.T. Cummings and C. McCabe, "A Wang-Laudau study of Bilayer formation on a lattice model," *J. Chem. Phys.* **137** (2012) 144901

5. L. Gai, T. Vogel, K. Maerzke, C. Iacovella, D.P. Landau, P.T. Cummings, and C. McCabe, "Examining the Phase transition behavior of amphiphilic lipids in solution using Statistical Temperature Molecular Dynamics and Replica-Exchange Wang-Landau Methods," *J. Chem. Phys.* submitted.

(329h) A Hybrid Fluctuating Hydrodynamics and Molecular Dynamics (Hybrid FHD/MD) Simulation Methods

Chu, J. W., University of California, Berkeley
Voulgarakis, N., University of California Berkeley
Originally presented on: 11/5/2013 14:36:00 - 14:54:00

The development of a hybrid FHD/MD simulation method that combines the molecular dynamics of moving particles with the fluctuating hydrodynamics of solvent fields on grid cells fixed in space is presented. This method allows resolution of solute-solvent interfaces and realization of excluded volumes of particles in the presence of hydrodynamic coupling. With these capabilities, we showed that the ubiquitous forces mediated by the solvent, hydrophobicity and hydrodynamics, can be linked in a mesoscopic simulation. The strategies we devised to overcome the numerical issues of mixing Eulerian and Lagrangian variables: using an auxiliary fluid to fill in the volumes excluded by particles and assigning collocating griding systems on solutes to interface with solvent fields, are also discussed. Simulation results show that the hybrid FHD/MD method can reproduce the solvation free energies and scaling laws of particles dynamics for hydrophobes of difference sizes. The collapse of two hydrophobic particles was also simulated to illustrate that the hybrid FHD/MD method has the potential to be generally applied to study nano-scale self-assembly and dynamics-struction-function relationships of biomolecules.

(384a) Efficient Property Prediction Using Expanded Ensemble Simulations Over Many Thermodynamic States

Shirts, M. R., University of Virginia
Kool, A. C., University of Virginia
Originally presented on: 11/5/2013 15:15:00 - 15:33:00

Expanded or generalized ensemble methods are a class of simulation methods that can be used to overcome sampling difficulties by adding transitions between thermodynamic ensembles. These additional thermodynamic ensembles can either be ensembles of interest (for example, the same system at different temperatures if temperature dependent effects are of interest) or auxiliary states added to either improve overlap between end states or to improve sampling by introducing states with faster kinetics.

One weakness of these methods is that in order to sample states at all ensembles with reasonable frequency, simulation weights for each of the subensembles must be introduced, which are equal to the free energies of each of the ensembles. These weights must be determined self-consistently, and the precise choice of algorithm to determine these self-consistent algorithms is an important choice in the implementation of these algorithms.

We show how approaches such as Wang-Landau and transition matrix methods can be generalized when dealing with expanded ensemble systems containing many thermodynamic states, and specifically as approximations to optimal multistate free energy estimators such as MBAR (the multistate Bennett's acceptance ratio method). This more general multistate approach allows more efficient determination of free energies and other observables over many states simultaneously, especially when the states are arranged in a multidimensional network, with multiple neighbors for each sate.

We additionally show how these multistate estimation methods can be combined with Gibbs sampling techniques, a generalization of Metropolis Monte Carlo methods which includes transitions to all possible states. This combination allows efficient sampling and property prediction over hundreds or thousands of thermodynamic states. We look at applications of these combined methods to searching the parameter space of spatially dependent Ising models as well as as well computing the binding free energy of small molecules to simple molecular containers.

(384b) Erroneous Reduction of Energy Barriers By One-Dimensional Potentials of Mean Force

Kopelevich, D. I., University of Florida

Originally presented on: 11/5/2013 15:33:00 - 15:51:00

A common approach to measurement of activation barriers for rare events is to enhance sampling of the system configurations in a neighborhood of the energy barrier. This can be accomplished by various methods ranging from umbrella sampling to forward flux simulations. Most of these methods yield the potential of mean force (PMF) acting on a key degree of freedom (reaction coordinate).

In this talk we discuss a possible pitfall associated with application of PMF parametrized by a single reaction coordinate to determine the height of the energy barrier. As a test case we consider transport of a hydrophobic nanoparticle into a lipid bilayer studied by coarse grained molecular dynamics (CGMD) simulations. The PMF acting on the nanoparticle and the nanoparticle diffusion coefficient are obtained from constrained MD simulations. Solution of the Langevin equation with thus obtained PMF and diffusivity predicts the mean transport time of 40 ns. This timescale is readily accessible by CGMD and therefore we can verify the prediction of the Langevin equation by a series of direct (unbiased and unconstrained) MD simulations. Surprisingly, the mean transport time observed in the direct simulations is 2 microseconds, i.e. almost two order of magnitude slower than that predicted by the one-dimensional Langevin model.

This discrepancy is caused by an inability of a one-dimensional PMF to capture the true energy barrier experienced by the nanoparticle. We show that the true energy barrier becomes hidden when one averages over all degrees of freedom transversal to the reaction coordinate. This apparent reduction of the energy barrier is caused by bistability of the system in a small neigborhood of the barrier. In order to determine the true barrier, it is necessary to account for an additional degree of freedom (in this case, the membrane deformation) and consider a multi-dimensional free energy landscape.

In conclusion of the talk, we present evidence that this situation is relatively common and discuss possible approaches to detection of such hidden energy barriers and modifications to the constrained MD simulations to determine the true barrier height.

(384c) Information Thermodynamics and Bayesian Optimization: Applications to Single Molecule Experiments

Haas, K. R., University of California, Berkeley
Chu, J. W., University of California, Berkeley
Yang, H.,
Originally presented on: 11/5/2013 15:51:00 - 16:09:00

We present a fundamental analysis of the information content for dynamics trajectories that is applicable to all systems governed by over-damped Langevin dynamics. The Fisher information metric for this class of time propagators is presented and the trajectory entropy functional is derived. This subsequently permits evaluation of the magnitude of descriptive features in the dynamic model parameterized by the potential of mean force and diffusion coefficient along a characteristic system coordinate(s).

Armed with a measure of "entropy" for a systems equilibrium trajectories, we are able to define a new manifold of information thermodynamics to balance the inputs of "internal energy" or log-likelihood statistical information from simulation and experiment with the desire to generate the simplest models of system behavior that contain only statistically significant features. This formulation permits the derivation of the least informative dynamic models subject to kinetic constraints such as mean-first passage times. Finally, we provide specific examples from performing Bayesian optimization of model parameters from single-molecule FRET experiments. Exploring what is the equivalent of a phase diagram for the operation of inverse mapping of data to models, allows for an optimal choice of regularizing parameters for model selection.

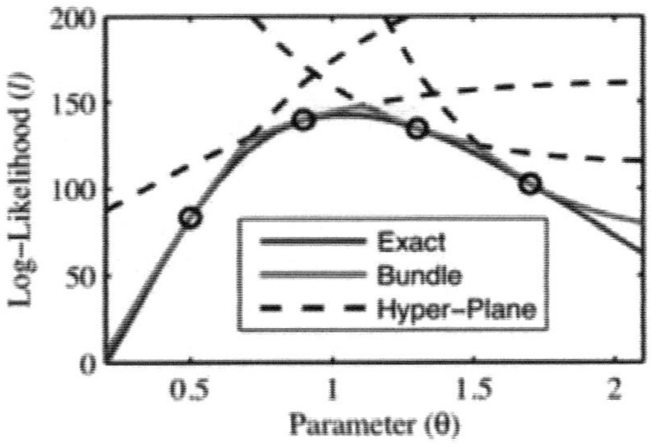

Figure 1. Functional approximation of the log-likelihood surface for a multi-dimensional parameter set θ which is a composite of the potential of mean force *V(x)* and the diffusion constant *D*. The dotted lines are hyper-planes in the functional space generated by projecting derivatives of the log-likelihood. Red lines are the convex hull created by the set of these upper bounds, and blue line is the true experimental/simulation likelihood as a parameterized by θ.

(384d) Grote-Hynes Theory, Recrossing, and Dynamics of the Committor in Ion-Pair Dissociation

Mullen, R. G., University of California
Shea, J. E., University of California
Peters, B., University of California Santa Barbara
Originally presented on: 11/5/2013 16:09:00 - 16:27:00

For chemical reactions in a solvent, friction emerges as a consequence of projecting multidimensional dynamics onto the dynamics of a single reaction coordinate. Friction causes recrossing of the dividing surface which can either be accounted for with a transmission coefficient or variationally minimized by optimizing the dividing surface. From a perfect non-recrossing dividing surface, all trajectories would commit to opposite basins in forward and backward time, transition state theory would become exact, the transmission coefficient would become one, and the committor distribution would become perfectly focused at ½. For Na^+Cl^- dissociation in TIP3P water, we show that recrossing persists even when the ½-committor surface itself is used as the dividing surface. We provide evidence that recrossing cannot be fully eliminated from the dynamics for any configurational coordinate. Consistent with this finding, inertial likelihood maximization finds a combination of ion-pair distance and two solvent coordinates that improves the committor distribution and increases the transmission coefficient relative to those for ion-pair distance alone, but recrossing is not entirely eliminated. Free energy surfaces for the coordinates identified by inertial likelihood maximization show that the solvent friction stems from anharmonicity and shallow intermediates that remain after dimensionality reduction to the dynamically important variables.

(384e) Markovian Milestoning for Computing Diffusion Rates of Ligands in Proteins

Lapelosa, M., Drexel University
Yu, T. Q., New York University
Vanden-Eijnden, E., New York University
Abrams, C. F., Drexel University
Originally presented on: 11/5/2013 16:27:00 - 16:45:00

Measuring diffusion rates of ligands plays a key role in understanding the kinetic processes inside proteins. For example, although many molecular simulation studies have reported free energy barriers to infer rates for CO diffusion in myoglobin (Mb), they typically do not include direct calculation of diffusion rates because of the long simulation times needed to infer these rates with statistical accuracy. We show in this talk how to apply Markovian milestoning along minimum free-energy pathways to calculate diffusion rates of CO inside Mb. In Markovian milestoning, one partitions a suitable reaction coordinate space into regions and performs restrained molecular dynamics in each region to accumulate kinetic statistics that, when assembled across regions, provides an estimate of the mean first-passage time between states. The mean escape time for CO directly from the so-called distal pocket (DP) through the histidine gate (HG) is estimated at about 24 ns, confirming the importance of this portal for CO. But Mb is known to contain several internal cavities, and cavity-to-cavity diffusion rates are also computed and used to build a complete kinetic network as a Markov state model. Within this framework, the effective mean time of escape to the solvent through HG increases to 30 ns. Our results suggest that carrier protein structure may have evolved under pressure to modulate dissolved gas release rates using a network of ligand-accessible cavities.

(384f) Enhanced Sampling Methods for Structural Refinement of Proteins and Nucleic Acids

Vashisth, H., University of Michigan
Brooks, C. L. III, University of Michigan

Originally presented on: 11/5/2013 16:45:00 - 17:03:00

Flexible-fitting computational algorithms are often useful in the interpretation of low-resolution electron microscopy (EM) data on macromolecular assemblies. Molecular dynamics flexible fitting (MDFF) is a widely used atomistic simulation technique to generate structural models of such biomolecular complexes. All-atom explicit-solvent MDFF simulations are not only computationally demanding, but also can be sensitive to the resolution of the target EM map. Moreover, functional movements of many biomolecules require large-scale conformational reorganization elicited via domain translations/rotations, where methods such as MDFF may be limited in capturing the rotations of structural elements. Therefore, novel enhanced sampling methodologies need to be integrated with MDFF to accelerate the conformational search in a single atomistic simulation. Such strategies have the potential to decrease the computational cost and alleviate the limitations stemming from conformational reorganization. In this work, we judiciously combine MDFF with temperature-accelerated molecular dynamics (TAMD), an enhanced sampling method, and carry out TAMD-assisted MDFF (TAMDFF) simulations of proteins and nucleic acids. We find that TAMDFF simulations can achieve target structures of similar quality as MDFF on shorter timescales. In some cases, only TAMDFF simulations are able to capture conformational changes likely because MDFF simulations are unable to overcome the underlying free-energy barriers. We suggest that TAMDFF may be a viable strategy for structural refinement of proteins, nucleic acids, and large ribonucleoprotein complexes such as the ribosome.

(384g) Metadynamics Simulations of Dissolution of Minerals. Overcoming Hidden Energy Barriers Problem

Hlushak, S., Vanderbilt University
Cummings, P. T., Oak Ridge National Laboratory
Originally presented on: 11/5/2013 17:03:00 - 17:21:00

Free energy profiles of the dissolution of silica from several flat crystallographic surfaces of α-quartz were obtained using a novel scheme based on metadynamics [1]. The multidimensional free energy profiles thus obtained are analyzed and compared to other experimental and simulation data. They suggest that the reaction coordinate of single collective variable is not sufficient to accurately sample activation energies between the intermediate dissolution states of the silica unit. The true activation barriers are hidden by the adjacent free-energy basins during the averaging/integrating over the excess collective variables. Consequently, the resulting one-dimensional free-energy profiles underestimate the height of the energy barriers. A space of three collective variables is demonstrated to be sufficient to reconstruct, with help of metadynamics, the free-energy profiles of silica unit dissolution and to more accurately predict free-energy barriers between intermediate states. We believe that the activation energies obtained from the multidimensional free energy profiles calculated using the proposed methodology should be more correct when compared to the other conventional approaches based on the one-dimensional blue moon ensemble and umbrella sampling methods. The metadynamics calculations were perfomed using both a classical reactive force field ReaxFF [2] and ab initio methods [3].

[1] A. Laio, F. L. Gervasio. Reports on Progress in Physics 71, 126601 (2008).
[2] J. C. Fogarty, H. M. Aktulga, A. Y. Grama, A. C. Van Duin, S. A. Pandit, The Journal of chemical physics 132, 174704 (2010)
[3] P. Giannozzi, S. Baroni, N. Bonini, M. Calandra, R. Car, C. Cavazzoni, D. Ceresoli et al. Journal of Physics: Condensed Matter 21, 395502 (2009).

(384h) Efficient Calculation of Solid-Fluid and Solid-Solid Coexistence, Including Consideration of Point Defects

Moustafa, S., University at Buffalo
Schultz, A. J., University at Buffalo, The State University of New York
Kofke, D. A., University at Buffalo, The State University of New York
Originally presented on: 11/5/2013 17:21:00 - 17:39:00

Thermodynamic stability of crystals is a topic of growing importance to a number of solid-state and nano-material applications of molecular modeling. This includes molecular crystals, (especially pharmaceuticals), alloys, refractory materials, clathrate hydrates, colloids, nanoparticles, and more. Some of these systems are amenable to the application of ab initio methods for calculating the molecular energy, and thus promise a first-principles approach to prediction of stable crystal structures. In other cases, the ``molecules" are models for nanoparticles, and the aim is to determine the structures that they will adopt via self-assembly. In any case, the crystal structure determines all other material properties, so knowledge of it for a given model is crucial to almost all materials applications. In many cases there are a large number of candidate structures that can be adopted at equilibrium, and identification of the most stable crystal structure is performed via reference to the thermodynamic free energy. Accordingly, it is important to have efficient methods to compute free energies for crystalline phases, and methods that work well for fluids often fail in this application.

In previous work [1, 2], we demonstrated that crystal free energies could be computed very efficiently using an interacting harmonic reference, following a path from low to high temperature, and using overlap sampling free-energy perturbation coupled with harmonically targeted perturbations in temperature. Such methods are inapplicable to hard potentials, which are of interest in modeling of nanoparticles and colloids. In the present work, we consider methods for such systems. We emphasize a quantitative comparison of the performance of the methods, using the hard sphere system as a prototype, and aiming to determine the most efficient scheme to evaluate the absolute free energy precisely in the shortest time. We find, for example, that the widely-used Frenkel-Ladd method is not the best approach to apply, and that other techniques can be as much as an order of magnitude more efficient in determining the free energy.

Another feature of phase equilibria of solids, and not seen in fluids, is the role of defects on behavior. Point defects are present in equilibrium solids, and can have an influence on properties disproportionate to their concentration. While methods have been developed to evaluate defect concentrations, they are cumbersome to apply. Here we present an approach that permits rapid evaluation of defect concentrations across a range of coexistence conditions. We demonstrate with the penetrable sphere model, which exhibits equilibrium between a fluid phase and a solid with vacancies, and between multiple-occupancy solids, and thus for which point defects are particularly important to the behavior.

References:

[1] T. B. Tan, A. J. Schultz, and D. A. Kofke, J. Chem. Phys. 133, 134104 (2010).
[2] T. B. Tan, A. J. Schultz, and D. A. Kofke, Journal of Chemical Physics 135, 044125 (2011).

(434a) Multiscale Modeling of the Shock Compression of Energetic Materials Using Particle Based Constant Energy Dissipative Particle Dynamics With Local Density Dependent Potentials

Moore, J. D., U.S. Army Research Laboratory
Izvekov, S., U.S. Army Research Laboratory
Lisal, M., Academy of Sciences of the Czech Republic
Brennan, J. K., U.S. Army Research Laboratory
Originally presented on: 11/6/2013 8:30:00 - 8:50:00

Energetic materials often contain nano- and microscale defects, such as voids and grain boundaries. Mechanical stimulation of these materials often incites responses over a wide range of spatial and temporal scales, with a strong dependence upon the defects. Modeling these materials atomistically remains a challenge due to the length and time scales required, as billions of molecules would be necessary to model micron-size defects. To overcome these challenges, we have implemented multiscale techniques to bridge the atomistic and mesoscale descriptions by coarse-graining RDX through a force-matching technique [1,2], resulting in a density-dependent potential which spans ambient to high pressures (>10 GPa) [3].

Our resulting model reasonably reproduces several atomistic properties [3,4], but only those properties which depend on intermolecular interactions. Properties that depend on the coarse-grained intramolecular degrees-of-freedom (e.g., the heat capacity) are underestimated [3]. Implementing traditional molecular dynamics to simulate the mechanical response of such models inevitably results in inaccurate energy and momentum exchange due to these unaccounted degrees of freedom. To correct this, we utilize the constant-energy Dissipative Particle Dynamics method (DPD-E) [5-7], which provides a mechanism to account for all coarsened degrees of freedom through the inclusion of a coarse-grain particle internal energy. This work presents results for the shock compression of RDX using DPD-E with results assessed by direct comparison to fully atomistic simulations. An additional focus will also be made concerning the derivation and implementation of the density-dependent potential, which will be shown to conserve energy.

[1] S. Izvekov and G.A. Voth, *J. Chem. Phys.*, **123**, 134105 (2005).

[2] S. Izvekov, P.W. Chung, and B.M. Rice, *J. Chem. Phys.*, **133**, 064109 (2010).

[3] S. Izvekov, P.W. Chung, and B.M. Rice, *J. Chem. Phys.*, **135,** 044112 (2011).

[4] S. Izvekov, P.W. Chung, and B.M. Rice, *Int. J. Heat and Mass Transfer*, **54**, 5623 (2011).

[5] J. Bonet Avalos and A.D. Mackie, *Europhys. Lett.*, **40**, 141 (1997).

[6] P. Español, *Europhys. Lett.*, **40**, 631 (1997).

[7] M. Lísal, J.K. Brennan, and J. Bonet Avalos, *J. Chem. Phys.*, **135**, 204105 (2011).

(434b) Toward Transitive Closure in Refinery-Wide Modeling: Atomically Explicit Kinetics Models

Bennett, C. A., University of Delaware
Hou, Z., University of Delaware
Zhang, L., China University of Petroleum
Klein, M. T., University of Delaware
Originally presented on: 11/6/2013 8:50:00 - 9:10:00

Toward Transitive Closure in Refinery-Wide Modeling: Atomically Explicit Kinetics Models

Craig A. Bennett[1], Zhen Hou[1], Linzhou Zhang[2], Michael T. Klein[1]

[1] *Department of Chemical and Biomolecular Engineering, University of Delaware, Newark, DE*

[2] *State Key Laboratory of Heavy Oil Processing, China University of Petroleum*

A refinery is largely made up of reactor units that work together to produce a series of usable end products from an initial stock of crude (or synthetic crude) oil. Although these reactor units are all tied together through a maze of pipelines, the models used to predict and evaluate the reactors have, for the most part, stood alone. Lumped models have largely been employed based on the needs of the reactor being modeled. As the product lumps of one reactor become the feed for a subsequent reactor, the lumping strategy largely has to change. Every time a lump is created and redistributed, a loss of information occurs. Therefore, a system of tracking and modeling the specific molecules involved would be preferable.

The in-house software tool known as INGen (the Interactive Network Generator) has the capability of compiling exhaustive kinetic networks for each reactor unit on an atomically explicit molecular-level. The reaction network is derived from an initial selection of feed molecules, allowed reaction rules, and a series of chemically based bookkeeping approximations. As was the case of the lumped model, the product stream of one reactor serves as the input stream to the next, but no information is lost in the molecular case as there is no translation from and to reactor specific lumps. As each successive reactor is likely to produce new chemical species, it is important to insure that every reactor has the ability to interface with the superset of all possible chemical species, the Universal Oil.

To assure this transitive closure, an initial oil composition consisting of species with five-aromatic rings or fewer (up to the boiling point range of a VGO) was passed through a series of individual reactor simulations: VGO Hydrocracking, Hydrotreating, FCC, Naphtha Hydrotreating, Cat Reforming, and Isom. The results of the end-of-line Isom unit were then used as the initial feed to the VGO Hydrocracking unit and the series of network generations was continued recursively until no new species were generated.

The resultant Universal Oil defines the common interface for the plant model units. Linear Programming and other efficiency modeling strategies built upon these models will be based on a level of detail and accuracy formerly unseen. Specific molecule tracking and monitoring will lead to better understanding of the processes and more efficient flow sheeting and design.

(434c) Rectangular Geometry of Catalytic Pore Domains for Electrochemical Systems: Role of the Deviations From "Ideality"

Okoye, N. H., Tennessee Technological University
Arce, P. E., Tennessee Technological University
Originally presented on: 11/6/2013 9:10:00 - 9:30:00

The study of catalytic pore in electrochemical systems is vital in determine the effectiveness region and acceptable pore size in chemical processes. This research outcome will be of help to predictions of system behaviors and assessing experimental results relevant to industries e.g. batteries, fuel cells. The "ideal" rectangular geometry for a case of, for example, anode catalyst layer of direct formic acid fuel cell is analyzed and compared with a geometry that deviates from the "ideal" strait one used to model the catalysts pore. This study will, for example, exemplify the role of the use of idealized geometrical domains (and their accuracy) in the prediction of system behaviors. A previous analysis (Nogalu and Arce, AIChE Annual Meeting, 2011) indicated that regions of parameter in the systems could lead to large errors for the sysem-level predictions. An effect in the deviation of the idealized geometry is still needed. This system is analyzed by employing the species mass balancing conservation equation and scaled up to the macroscopic level by employing area averaging-based approach. The system is solved analytically in conjunction with the scaling up by using the area averaging approach. The results obtained from the area averaging will enable us to determine the accuracy in, for example, the effectiveness factor for the system. The deviation from "ideality" will be quantified and the error percentage noted. Suggestions for the practical implications will be offered and future work, outlined.

(434d) Modelling of Varying Length-Scaled Catalytic Films in Electrochemically Promoted Systems

Fragkopoulos, I. S., University of Manchester
Theodoropoulos, C., University of Manchester
Originally presented on: 11/6/2013 9:30:00 - 9:50:00

Electrochemical Promotion of Catalysis (EPOC) is the enhancement of the catalytic activity when potential or current is applied in systems where a metal catalyst is deposited on a solid electrolyte. Due to this application, a symbloc of species [$O^{\delta-}$ - $\delta+$] known as 'backspillover' species are generated at the Triple Phase Boundaries (TPBs, places where the electrolyte, the gas phase and the metal meet) and spill over the catalytic film producing an effective double layer. It has been found that the effective double layer affects the binding strength of catalytic surface adsorbates, resulting in a dramatic enhancement of the catalytic activity [1,2]. EPOC effect also referred to as Non-Faradaic Electrochemical Modification of Catalytic Activity (NEMCA) was found that can also be extended to oxide catalysts, whose activity can be increased by up to a factor of 10 via anodic polarization [3]. Although EPOC exhibits a great potential nowadays in the fields of modern electrochemistry [4,5], it has still not found industrial applications, mainly because the underlying phenomenon is not fully understood yet and cannot be modelled to allow robust system design [6] and control.

In this work we first study the behaviour of supported catalysts of varying nano-structured length scales under open circuit and high pressure conditions in order to get a better understanding on the catalyst-support boundary and its effect on the catalytic activity and the surface reaction rates. We propose an in-house developed efficient implementation of the kinetic Monte Carlo method (kMC) [7] for the simulation of reaction-diffusion micro-processes on the catalytic films and in conjunction with relevant experimental data (Prof Ian Metcalfe's group) we obtain insights of this open circuit "enhancement" phenomenon as well as reliable estimates of the surface reaction kinetics.

We subsequently exploit the obtained kinetic parameters to construct an accurate multi-scale framework [8] for the same nano-structured systems to quantify the Electrochemical Promotion effect under closed circuit conditions. The multi-scale framework couples a 3-D macroscopic model based on partial differential equations (PDEs) simulating charge transport as well as electrochemical phenomena at the TPBs implemented through commercial software (COMSOL Multiphysics) and the 2-D kMC simulator described above. Dynamic coupling of macro- and microscopic models at the TPBs leads to the construction of an integrated multi-scale system. Moreover, due to computationally extensive 'large' kMC simulations, the catalytic surface is split into a number of smaller "representative" lattices whose total area is only a fraction of the actual catalytic area and intelligent interpolation techniques [9,10] are employed to simulate the interactions between the lattices including lateral (lattice-to-lattice) transport through diffusion. The system considered in this study comprises of a thin Pt catalytic film on an YSZ support and Au reference/counter electrodes for the oxidation of carbon monoxide.

References

1. M. Stoukides, C.G. Vayenas, J. Catal. 1981; 70(1): 137-146.

2. C.G. Vayenas, S. Bebelis. S. Ladas, Nature. 1990; 343: 625-627.

3. J. Nicole, C. Comninellis, Solid State Ionics. 2000; 136-137: 687-692.

4. D. Poulidi, M.E. Rivas, I.S. Metcalfe, J. Catal. 2011; 281(1): 188-197.

5. E.P.M. Leiva, C.Vazquez, M.I.Rojas, M.M. Mariscal. J. Appl. Electrochem. 2008; 38(8): 1065-1073.

6. N.A. Anastasijevic, Catal. Today. 2009; 146:308-311.

7. S. Raimondeau, D. G. Vlachos, Comput. Chem. Eng. 2002; 26(7-8), 965-980.

8. I.S. Fragkopoulos, I. Bonis, C. Theodoropoulos, Computer-Aided Chem. Eng. 2012; 30: 1013-1017.

9. C.W. Gear, J. Li, I.G. Kevrekidis, Phys. Lett. A. 2003; 316: 190-195.

10. A. Armaou, I.G. Kevrekidis, C. Theodoropoulos, Comput. Chem. Eng. 2005; 29 (4): 731-740.

(434e) Modeling and Analysis of Coupled NOx Storage and Reduction and Selective Catalytic Reduction Monolithic Catalysts

Shakya, B. M., University of Houston
Harold, M. P., University of Houston
Balakotaiah, V., University of Houston
Originally presented on: 11/6/2013 9:50:00 - 10:10:00

Modeling and Analysis of Coupled NOx Storage and Reduction and Selective Catalytic Reduction Monolithic Catalysts

Bijesh M. Shakya, Michael P. Harold[1], and Vemuri Balakotaiah[2]

Department of Chemical & Biomolecular Engineering

University of Houston 77204-4004

Abstract

The combination of NOx storage and reduction (NSR) and selective catalytic reduction (SCR) catalyst has emerged as a promising technology for the reduction of NOx emission from the exhaust of lean-burn or diesel engine vehicles. NSR is operated in periodic fuel-lean and fuel-rich cycles in which NOx is stored on the storage component of the catalyst, also called a lean NOx trap (LNT), during the lean-mode and is reduced during the rich-mode. It is widely accepted that ammonia is formed as an intermediate during the regeneration of stored NOx which further reacts to form N_2. On the other hand, SCR technology uses transition metal (Cu, Fe) exchanged zeolite catalyst and utilizes NH_3 generated from external source to reduce the NOx from the lean exhaust. In the combined NSR/SCR system, NH_3 generated in LNT during the rich phase is stored in the SCR which is used to reduce the NOx during the subsequent lean phase. In this way, the integrated system not only eliminates the need to have external source of NH_3, as required in standalone SCR system, but also provides an opportunity to lower the volume of expensive PGM based LNT catalyst, both of which lower the overall cost of the aftertreatment system.

Many experimental and modeling works have been done addressing various aspects of the combined system [1, 2]. Notwithstanding these recent developments, a detailed modeling study focusing on the dual-layer LNT/SCR configuration is still lacking. In this study, we developed a new global kinetic model for low Pt dispersion (8%) LNT and used it in conjunction with previously developed kinetic model for Cu-Chabazite SCR catalyst [3] to perform simulations of combined LNT/SCR catalyst [4]. Specifically, we analyzed the impact of washcoat loading of individual component and effect of temperature in the dual-layer architecture. These results were then compared with the brick configuration under identical operating conditions and catalyst composition (i.e. same fraction of LNT and SCR). Fig 1 shows the model predicted cycle-averaged NOx conversion and N_2 selectivity as a function of SCR fraction in dual-layer and dual-brick LNT/SCR catalyst at 230°C and 300°C. The simulation results show that NOx conversion attains a maximum as the SCR loading is increased in dual-layer catalysts. At higher SCR loading, the number of adsorption sites in SCR is much larger than the amount of NH_3 that is generated in LNT and as a result only a fraction of SCR layer close to LNT are utilized while the rest act as an inert layer or a diffusional barrier thereby lowering the conversion. Such a diffusional barrier is not significant for the case of brick configuration at these temperatures (230°C and 300°C). Fig 1 also shows that maximum NOx conversion is obtained by dual-layer configuration at both temperatures. This is due to the improved utilization of in-situ generated NH_3 on the SCR layer which can be clearly seen by the relatively higher selectivity of N_2 obtained in dual layer catalyst compared to that of dual brick. However, at higher loading of SCR, NOx conversion obtained by dual-layer catalyst is lower than that by dual brick for the reasons mentioned above.

(434f) Multiscale Description of Transport and Reaction in Coated Catalyst Layers

Marek, M. A., Prague Institute of Chemical Technology

Originally presented on: 11/6/2013 10:10:00 - 10:30:00

Multiscale description of transport and reaction in coated catalyst layers

Vladimir Novak[a], Petr Koei[a,*], Toma¡? Gregor[b],

Franti?ek ?tÃ¬pÃ¡nek[a], Milo? Marek[a]

[a] Department of Chemical Engineering, Institute of Chemical Technology, Prague, TechnickÃ¡ 5, CZ 16628 Prague, Czech Republic

[b] New Technologies Research Centre, University of West Bohemia, UniverzitnÃ 8, CZ 30614 Pilsen, Czech Republic

([*]Corresponding Author's E-mail: milos.marek@vscht.cz)

ABSTRACT

The coated catalytic layer In standard 1D+1D models of monolith channel, is usually treated as homogeneous, with transport properties represented by effective diffusivity [1]. This effective diffusivity is typically calculated from random pore model [2] that considers contribution of both small micro/mesopores (intrinsic nanostructure of the used material) and larger macropores (inter-particle structure influenced by the coating procedure). In reality, however, the coated layer may contain additional cracks and cavities (Fig. 1). These structural defects can enhance effective diffusivity in the layer and improve the performance of the catalyst, so that their occurrence is beneficial as long as the adhesion and mechanical durability of the layer is preserved.

In this contribution we apply novel experimental and computational methods to investigate the contribution of cracks and cavities to diffusion in catalytic coatings. A sample of commercial catalytic monolith for automotive exhaust gas aftertreatment is scanned by X-ray computer microtomograph (μCT) to obtain realistic 3D images of the coated layer (Fig. 1). Spatially 3D diffusion is then simulated within a section of the reconstructed porous structure, using a validated methodology [3,4,5,6]. Additional simulations are performed for the model coating without any cracks or cavities (i.e., containing only standard meso- and macro-pores). By comparison of the simulation results, the contribution of cracks and cavities to transport characteristics of the layer is quantified. The evaluated effective diffusivities are compared with random-pore model approximations [2], and further utilized in multi-scale simulations of combined reaction and transport [3] that reveal the impact of transport effect on overall performance of the monolith reactor.

The proposed combination of advanced 3D structural analysis technique and detailed modeling approach significantly extends the level of detail in understanding the internal transport effects in catalytic coatings of monoliths.

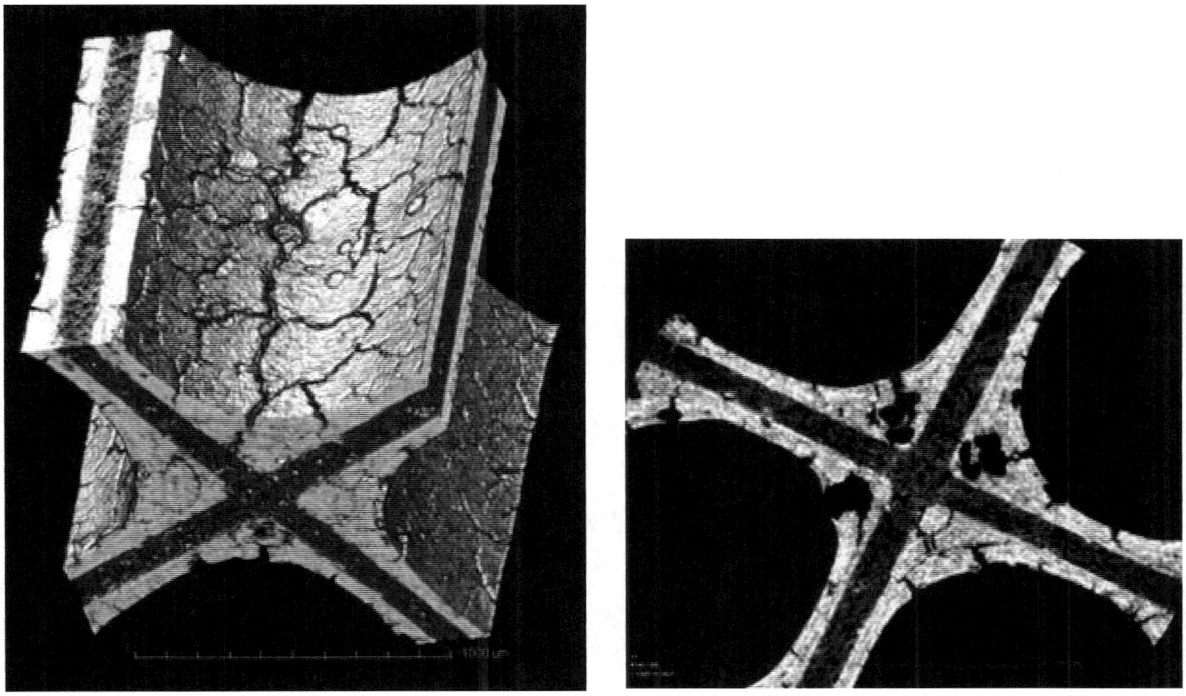

Fig. 1. 3D reconstruction of catalytic coating in commercial monolith with square channels (CLEERS research community [7] reference commercial catalyst LNT BMW 120i model year 2009) ? sections around the channel corner are shown. Monolith substrate is displayed as dark grey, catalytic coating is bright, and black color corresponds to void space (pores, cavities and cracks). The sample scanned by X-ray computer microtomograph (μCT).

Acknowledgements

This work has been financially supported by the Czech Science Foundation (GACR P106/10/1568). The X-ray microtomography scans and 3D reconstruction results were obtained within the CENTEM project CZ.1.05/2.1.00/03.0088, co-funded by the European Research Development Fund as part of the Czech Ministry of Education, Youth and Sports' OP RDI programme. The work was further supported by the project LH 12086 of the Czech Ministry of Education, Youth and Sports.

References

[1] GÃ¼thenke A., Chatterjee D., Weibel M., Krutzsch B., KoÄ¨Ã P., Marek M., Nova I., Tronconi E., Current status of modelling lean exhaust gas aftertreatment catalysts, *Advances in Chemical Engineering* **33** (2007), 103-211.

[2] Wakao N., Smith J.M., Diffusion in catalyst pellets, *Chemical Engineering Science* **17** (1962), 825-834.

[3] KoÄ¨Ã P., NovÃ¡k V., ?tÄ›pÃ¡nek F., Marek M., KubÃÄ¨ek M. Multi-scale modelling of reaction and

transport in porous catalysts. *Chemical Engineering Science* **65** (2010), 412-419.

[4] NovÃ¡k V., ?tÃ¬pÃ¡nek F., KoÄ Ã P., Marek M. Evaluation of local pore sizes and transport properties in porous catalysts. *Chemical Engineering Science* **65** (2010), 2352-2360.

[5] NovÃ¡k V., KoÄ Ã P., Marek M., ?tÃ¬pÃ¡nek F., Blanco-GarcÃa P., Jones G. Multi-scale modelling and measurements of diffusion through porous catalytic coatings: An application to exhaust gas oxidation. *Catalysis Today* **188** (2012), 62-69.

[6] http://www.vscht.cz/monolith

[7] http://www.cleers.org

(434g) Modeling Tube-Scale Fixed Bed Radial Heat Transfer From Sub-Particle-Scale Velocity Components

Behnam, M., Worcester Polytechnic Institute
Dixon, A. G., Worcester Polytechnic Institute
Nijemeisland, M., Johnson Matthey Catalysts
Stitt, E. H., Johnson Matthey Catalysts
Originally presented on: 11/6/2013 10:30:00 - 10:50:00

Heat transfer in low tube-to-particle diameter ratio (N) fixed beds is important as these devices are extensively used as reactors, in applications such as steam reforming, partial oxidation and hydrogenation. Current approaches to radial fixed bed heat transfer modeling all suffer from serious deficiencies, as they average out the small-scale velocity components in the bed, replacing them with either plug flow or grossly simplified velocity models. Transport processes are then based on the classical k_r ? h_w effective parameter model or its extensions, in which transport mechanisms are represented only on an averaged tube diameter scale. A CFD-based multi-scale approach is proposed to developing a radial heat transfer model for fixed beds. In this approach we do not use the effective thermal conductivity k_r for thermal dispersion, a purely fluid mechanical phenomenon, or the apparent wall heat transfer coefficient $h_{w.}$ We have developed a new velocity-based approach that depends on the use of computational fluid dynamics (CFD) to obtain 3D flow fields on a sub-particle level; these are carefully coarse-grained to retain the essential features and to provide the tube-scale input to a new 2D model. CFD here also provided validated temperature profiles to test the new model.

Computer-generated beds of spheres were obtained through a modified soft-sphere collective rearrangement algorithm, for N = 3.96, 5.96 and 7.99. Overall, the features and magnitude of experimental radial void fraction profiles and bulk voidage were well-reproduced. CFD simulations of flow and heat transfer were carried out using the commercial code Fluent$^{Â®}$ version 6.3.26, for the values of N and for four values of Re in the range 80 - 1900. The simulations for lower Re were run as laminar (DNS) models, while the Re = 1900 simulations were run as turbulent k-ε models. The spherical particles were nominally 0.0254 m in diameter, reduced to 99% to avoid meshing problems. The mesh had boundary layer prism cells on the tube wall, with tetrahedral cells in the main fluid volume. Mesh refinement studies and validation against experimental data were done and have been reported previously.

The conduction contribution to the heat transfer model was obtained with a locally-varying stagnant thermal conductivity, $k_e^0(r)$ from the Zehner-SchlÃ¼nder cell model, as a function of true fluid thermal conductivity, particle thermal conductivity and the radial bed voidage profile. Our earlier work on conduction has validated this pointwise use of the Zehner-SchlÃ¼nder formula, which was originally developed to predict the overall, or bed-scale thermal conductivity.

For the convective contribution, we postulated that the dispersion of heat by small-scale radial displacements of flow could be represented by a bed-scale two-dimensional velocity field using the averaged components $v_z(r)$ and $v_r(r,z)$. The new 2D model equation is (for the pseudohomogeneous case):

$$\rho_f C_{pf}\left(v_r(r,z)\frac{\partial T}{\partial r} + v_z(r)\frac{\partial T}{\partial z}\right) = \frac{1}{r}\frac{\partial}{\partial r}\left(r\left(k_e^o(r)\right)\frac{\partial T}{\partial r}\right)$$

The boundary conditions are $T = T_{in}$ at $z = 0$, $T = T_w$ at $r = R$, and symmetry at the tube centerline $r = 0$. COMSOLÂ® finite element software was used to solve the pseudo-continuum 2D pseudohomogeneous heat transfer model. The pointwise velocities were implemented by the incorporation of tables for void fraction and axial velocity as functions of r, and radial velocity as a function of r and z, which were then interpolated as input to the finite element method.

The comparisons between the averaged CFD 3D discrete particle temperatures and the new 2D model temperatures showed excellent quantitative agreement. The new model predicted the axial temperature distribution fairly well; in addition the radial temperature distribution was predicted very well. The temperatures of the fixed beds had rougher distributions in the axial direction due to the particle heat transfer by conduction and the radial velocities distribution between particles. Both models illustrated development of the temperature into the center of the bed at the same locations.

(495a) Invited Talk: Using Molecular Simulations to Design Cationic Polymers for DNA Binding and Delivery

Jayaraman, A., University of Colorado, Boulder
Elder, R. M., University of Colorado at Boulder
Johnson, D.,
Originally presented on: 11/6/2013 12:30:00 - 13:00:00

Gene delivery involves successful transfection of therapeutic DNA by a vector into target cells and expression of that genetic material. Viral vectors, while effective, can elicit harmful immunogenic responses, thus motivating ongoing research on non-viral transfection agents. Cationic polymers or polycations are a promising class of non-viral vectors due to their low immugenic responses and low toxicity, and their ability to bind to the polyanionic DNA backbone to form a polycation-DNA complex (polyplex) that is then internalized in the target cell. Combinatorial approaches have generated many polycations with differing DNA transfection efficacies, but there is a need for general design guidelines that can relate the molecular features of the polycation to its DNA transfection efficiency. Using atomistic and coarse-grained molecular dynamics simulations of DNA and lysine based polycations with and without the nuclear localization sequence (PKKKRKV), we connect the thermodynamics of polycation-DNA binding to experimentally observed transfection efficiency of these polycations. In particular, our computational studies provide enthalpic and entropic contributions to DNA binding free energy, and the structure of the polycation-DNA complexes as a function of polycation chemistry and architecture.

(495b) Experimental Characterization and MD Simulation of the Steric Shielding Effect of Polyethylene Glycol (PEG) On the Conjugated Protein

Yu, J., Institute of Process Engineering, Chinese Academy of Sciences
Originally presented on: 11/6/2013 13:00:00 - 13:20:00

PEGylation, conjugation of polyethylene glycol (PEG) to a therapeutic protein, is a successful approach to improve potency of a therapeutic protein. The improved therapeutic potency is mainly due to the steric shielding effect of PEG. However, the underlying mechanism of this effect on the protein is not well understood, especially on the protein interaction with its high molecular weight substrate or receptor. Here, experimental study and molecular dynamics simulation were used to provide molecular insight into the interaction between the PEGylated protein and its receptor. Staphylokinase (Sak), a therapeutic protein for coronary thrombolysis, was used as a model protein. Four PEGylated Saks were prepared by site-specific conjugation of 5 kDa/20 kDa PEG to N-terminus and C-terminus of Sak, respectively. Experimental study suggests that the native conformation of Sak is essentially not altered by PEGylation. In contrast, the bioactivity, the hydrodynamic volume and the molecular symmetric shape of the PEGylated Sak are altered and dependent on the PEG chain length and the PEGylation site. Molecular modeling of the PEGylated Saks suggests that the PEG chain remains highly flexible and can form a distinctive hydrated layer, thereby resulting in the steric shielding effect of PEG. Docking analyses indicate that the binding affinity of Sak to its receptor is dependent on the PEG chain length and the PEGylation site. Computational simulation results explain experimental data well. The study clarified molecular details of PEG chain on protein surface. Similar studies can facilitate rational design, fabrication and clinical applications of PEGylated protein drugs.

(495c) Molecular Simulations of Compressibility of Co-Polymer Brushes

Panwar, A. S., Indian Institute of Technology Bombay
Chakrabarti, B., Durham University
Mishra, A., Indian Institute of Technology Bombay
Originally presented on: 11/6/2013 13:20:00 - 13:40:00

Responsive polymer brushes are potential candidates for a variety of applications in molecular sensing and selective-gating in nano-structured materials. Responsive brush-like structures are also found in biology, in the cases of cytoskeletons of neurons and nuclear-pore complexes. In this work, we investigate the compressibility of end-tethered co-polymer brushes, composed of hydrophilic and hydrophobic segments, in solution using Langevin dynamcis simulations. Primarily, the simulations address the influence of three major parameters on brush compressibility, namely, the grafting density, the fraction of hydrophobic groups in a co-polymer, and the relative distributions of hydrophobic and hydrophilic groups along the polymer contour. We consider the following distributions of segments for any given fraction of hydrophobic groups, periodically spaced segments, randomly distributed segments, and block co-polymer arragement, with respective hydrophilic and hydrophobic blocks. The equilibrium heights of these brushes show a strong dependence on both fraction of hydrophobic groups and their distributions along the polymer contour. The first observation is that the brush height decreases in a sigmoidal manner for both periodic and random distributions with increase in hydrophobic fraction. Moreover, the equilibrium height values are nearly the same for both periodic and random distributions. Further, we study the mechanical response of these brushes when the brushes are compressed in a reversible manner with a non-interacting wall. Again, we find that the bulk moduli of the co-polymer brushes show a strong dependence on the fraction and distribution of hydrophobic groups. The bulk modulus increases monotonically with hydrophobic fraction. However, the bulk modulus variation shows two distinct regimes with respect to increase in hydrophobic fraction, a low slope regime for smaller hydrophobic fractions, and a high slope regime for larger hydrophobic fractions. We discuss these results with respect to decreased segmental diffusivity and an increase in inter-segment interactions in the co-polymer brush.

(495d) Dynamics of Semi-Flexible Polymers in a Poor Solvent

Kong, M.,
Saha Dalal, I., University of Michigan
Larson, R. G., University of Michigan
Originally presented on: 11/6/2013 13:40:00 - 14:00:00

For polymers under poor solvent conditions, polymer chains form a compact globule, and the shape of the globule and conformation of the chain are sensitive to chain stiffness. While multiple studies of the dynamics and equilibrium conformations of such chains have been undertaken recently, to date no method of systematic coarse-graining such chains has been proposed. In this study, we explore coarse-graining procedures for investigating the conformational and dynamic behavior of polymer chains with bending stiffness in poor solvent conditions using Brownian dynamics (BD) simulations. We study chains with fixed Kuhn length, but varying numbers of rods representing each Kuhn length, and seek conditions under which both static shear-flow properties of the collapsed configurations become insensitive to the degree of coarse-graining. We find conditions of solvent quality and chain stiffness under which coarse-graining leads to converged results, and where coarse-graining breaks down due to strong local interactions. This method allows us to obtain general results for the effect of bending stiffness on the folding and unfolding of polymer chains under poor solvent conditions that are robust to the degree of chain resolution. We show that as chain stiffness increases, the collapse transition becomes steeper as solvent quality worsens, as does the unraveling transition that occurs under shearing flow.

(495e) Advanced Kinetic Monte Carlo Optimization of the Synthesis of Well-Defined (co)Polymers Via Atom Transfer Radical and Nitroxide Mediated Polymerization

Van Steenberge, P. H. M., Ghent University
D'hooge, D. R.,
Toloza Porras, C.,
Fierens, S. K.,
Reyniers, M. F., Ghent University
Marin, G. B., Ghent University
Originally presented on: 11/6/2013 14:00:00 - 14:20:00

In the last two decades, the kinetic Monte Carlo (kMC) technique has emerged as a powerful tool for the simulation and optimization of polymerization processes.[1-5] Benefiting from its natural sampling of reaction events, the polymeric microstructure can be simulated in molecular detail. Not only the chain length of the different macrospecies can be predicted but other important characteristics such as their topology, overall comonomer content and incorporated functionalities can be described too. Moreover, the recent evolution of advanced computer technologies has further increased the potential of the kMC technique by opening up the possibility to track the evolution of the complete polymeric microstructure during the polymerization.

Based on the work of Szymanski[5] on living polymerization and the work of Wang and Broadbelt[3,4] on nitroxide mediated polymerization (NMP), we have developed a general kMC methodology allowing to track the history of sampled reaction events for a representative group of macromolecules. For example, tracking kMC propagation events yields the monomer sequences of every macromolecule in this group throughout the polymerization process.[6,7] Such detailed information enables the calculation of new molecular properties *e.g.* macromolecular deviations from ideal targeted monomer sequences. In particular, for copolymers gradient (<GD>) and block deviation (<BD>) criteria were introduced to characterize, respectively, their linear gradient en diblock quality using atom transfer radical polymerization (ATRP) as polymerization technique. The main merit of the kMC methodology is its objective evaluation of the structural properties of synthesized microstructures, paving the way to unambiguously rank different copolymer products according to their quality. The latter is a challenging task in view of experimental difficulties encountered in the characterization of the polymeric microstructure in molecular detail.[8]

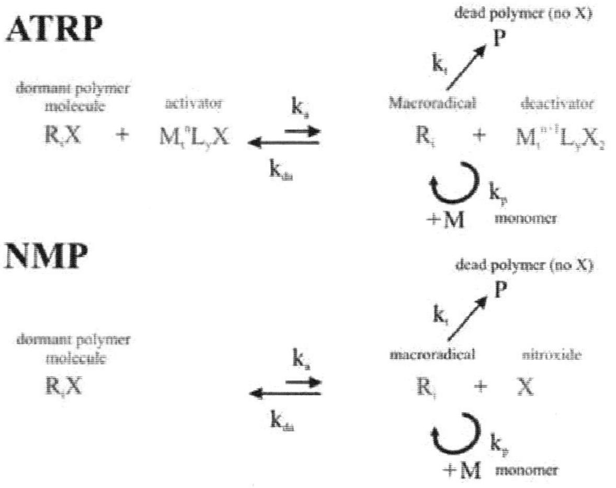

Scheme 1: Main reactions in (a) atom transfer radical polymerization (ATRP) (b) nitroxide mediated polymerization (NMP); R_0X (i=0): ATRP/NMP initiator (halogenide/alkoxyamine); i: chain

length; M_tL_yX: ATRP catalyst in low oxidation state (activator); M_t: transition metal; L: ligand; X: halogen atom/nitroxide: end-group functionality (EGF); P: dead polymer; R_iX (i>0): dormant polymer molecule; ideally no P formed and all dormant polymer molecules possess equal chain length.

The main reactions occurring in ATRP and NMP, two important controlled radical polymerization (CRP) techniques, are depicted in Scheme **1**. Initiator radicals (R_0; $i=0$) are generated from ATRP/NMP initiator molecules (R_0X). In ATRP, this activation requires the presence of an ATRP catalyst (M_tL_yX; activator) as mediating agent whereas in NMP a sufficiently high polymerization temperature is used to create the mediating nitroxide (X) by cleavage of the labile C-ON bond of an alkoxyamine initiator. The formed R_0 species ideally take up only a few monomer units before they are temporarily deactivated to a dormant form (R_iX) thereby introducing end-group functionality (EGF; X) in the polymeric species. For high deactivation rates, a low radical concentration is established ensuring a high livingness, *i.e.* the formation of dead polymer molecules with concomitant loss of EGF is reduced. Also, for fast ATRP/NMP initiation, dormant species with almost uniform chain length can be obtained at sufficiently high conversion, *i.e.* tailored polymer molecules both with respect to livingness and chain length result for CRPs after careful selection of the polymerization conditions and mediating agent.

Figure **1** illustrates the added value of the <GD> calculation in CRP processes. For the batch ATRP of methyl methacrylate (MMA) and *n*-butyl acrylate (*n*BuA), of MMA and styrene (STY) and of STY and nBuA, it has been shown that the <GD> at complete monomer consumption correlates with the dispersity (also known as polydispersity index).[9] For the three copolymerizations studied, the <GD>decreases for decreasing dispersity indicating that a good control over chain length is needed for optimal gradient polymer synthesis. The highest linear gradient quality is obtained for MMA/nBuA, as can be expected based on the inherent preference of the secondary ?nBuA radicals? to propagate with MMA with formation of tertiary ?MMA radicals?. Additionally, a fed-batch approach allows to improve the linear gradient quality in agreement with earlier deterministic simulations[10] at the expense of polymerization rate. Hence, the optimal addition profile can be predicted by means of kMC simulations, taking into account practical limitations.

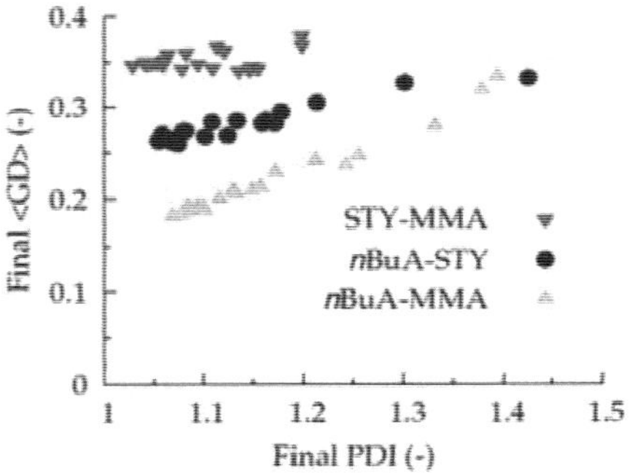

Figure 1: Correlation between linear gradient quality (expressed via <GD>) and dispersity at complete monomer consumption for the batch ATRP of different comonomer pairs A/B; common polymerization conditions: $[A+B]_0:[R_0X]_0:[C]_0 = 100:1:1$ with C the ATRP catalyst (only low oxidation state); $[A]_0=[B]_0$; perfect linear gradient for <GD> of 0; homopolymer: maximal <GD> of 1.[10]

In the present contribution it will be shown that the kMC methodology can also be used to identify ATRP/NMP conditions that allow to achieve a maximal control over chain length and EGF for the synthesis of well-defined poly(STY-*b*-acrylate) block copolymers taking into account constraints with respect to the maximal overall polymerization time and desired number average chain length (50-1000). For the copolymerization, a distinction is made between ?one-pot? and ?two-pot? approaches. In the ?two-pot? approach, the first block is, after purification, used as macro-initiator for the synthesis of the second block. In the ?one-pot? approach, the second monomer is added at high conversion of the first monomer and no extra purification step is performed. Hence, once the second monomer is added, both monomers can be incorporated in the macrospecies resulting in a block-like copolymer. To evaluate the effect of diffusional limitations on reaction steps with a high intrinsic reactivity, the concept of apparent rate coefficients[11] is applied.

For the ATRP process, focus is on the use of low catalyst amounts starting from the oxygen-stable deactivator in the presence of conventional radical initiator. It has been Â shown that a lower <BD> can be obtained with a ?two-pot? approach under otherwise similar conditions and that an optimal polymerization temperature exists for each block.[6] For the NMP process, commercially available alkoxyamine initiators are considered. Arrhenius parameters for nitroxide related reaction steps are determined from an extensive set of experimental data for NMP of both the first (poly(STY)) and the second block (poly(nBuA)). In the kMC model, a detailed reaction scheme including thermal[12] self-initiation of styrene and backbiting of acrylate[14-16] radicals is implemented to maximize the application field of the developed kMC model. It is illustrated that the alkoxyamine MAMA-SG1 (also known as Blockbuilder™) allows the production of well-tailored poly(STY-*b*-nBuA) block copolymers in case an appropriate temperature profile is applied for both the ?one-pot? and ?two-pot? approach.

(495f) Multiscale Modeling of Polyisoprene On Silica and Graphite

Brayton, A., University of Houston
Pandey, Y. N., University of Houston
Papakonstantopoulos, G. J., The Goodyear Tire & Rubber Company
Burkhart, C., The Goodyear Tire & Rubber Company
Doxastakis, M., University of Houston
Originally presented on: 11/6/2013 14:20:00 - 14:40:00

Materials created by dispersing nanoparticles in a polymer matrix strive to meet the promise of enhanced and often unique properties at a reduced cost. The availability of structure-property relationships and predictive modeling are deemed necessary to tailor these materials according to our needs. Molecular dynamics simulations of polymer-particle mixtures in atomistic detail is limited by exceedingly large systems required to be probed over the longest-relaxation time of the polymer on the surface.

In this study we examine the ability of systematically derived-coarse grain models (CG) to probe the configurations of *cis*-1-4 polyisoprene (PI) in proximity to silica and graphite surfaces. All atom (AA) molecular dynamic simulations of large systems are performed and details of chain configurations are extracted. Using the Iterative Boltzmann Inversion method[1] we construct interaction potentials between CG segments as well as between polymer segments and flat surfaces. We find that in the melt state, CG models can quantitatively describe the structure of the melt as verified by detailed comparison with AA models. We propose a scheme to extract interaction potentials between a nanoparticle and PI based on our calculations with flat surfaces. We find that with our approach we can capture accurately the melt structure around curved nanoparticles without the need of further optimization of the interaction parameters. Subsequently, we employ recently developed Monte Carlo algorithms that couple connectivity-altering moves with preferential sampling.[2] Our approach allows the accurate characterization of the polymer layer surrounding nanoparticles without current limitations imposed by chain-connectivity. Finally we discuss extensions of our work on extracting the interactions between two particles immersed in a polyisoprene matrix.

1. D. Reith, M. Putz, F. J. Müller-Plathe, *J. Comput. Chem.,* **24**, 1624–1636, 2003.

2. Y. Pandey and M. Doxastakis, , *J. Chem. Phys.*, **136**, 094901, 2012.

(495g) Coarse Grained Molecular Dynamics Model of Block Copolymer Directed Self-Assembly

Lawson, R. A., Georgia Institute of Technology
Peters, A., Georgia Institute of Technology
Ludovice, P. J., Georgia Institute of Technology
Henderson, C. L., Georgia Institute of Technology
Originally presented on: 11/6/2013 14:40:00 - 15:00:00

Directed self-assembly (DSA) of block copolymers (BCPs) is a promising technique for producing sub-30 nm pitch regular patterns, and the development of these DSA techniques could benefit greatly from computer simulation of such methods. Current simulation methods such as mean field approaches suffer from a number of limitations that affect their accuracy and their level of detail. Unlike other models, molecular dynamics (MD) combined with realistic potentials for polymer behavior can potentially provide more accurate simulations of the inherent polymer behavior, dynamics, and equilibrium states without a need to guess modes of molecular movement and without oversimplifying interatomic interactions. This is especially important for polymer systems that have very different densities or interaction strengths. We have developed a novel model for simulation of BCP behavior and DSA processes based on molecular dynamics of coarse-grained polymer chains simulated using graphics processing units (GPUs) to perform the calculations. The polymers consist of a bead-spring type system where a single bead corresponds to multiple monomers. Each bead undergoes 3 interactions: a harmonic spring-like bonded potential, a harmonic angle (consisting of 3 consecutive bonded beads), and a non-bonded potential modeled using a Lennard-Jones like potential. This coarse graining can result in simulations ~100-200 times faster than atomistic simulations, depending on what scale of coarse graining is used in the polymer model. In addition to this, this work uses graphics processing units (GPUs) to perform the calculations. This is very effective because of the natural parallelizability of MD simulations. Using HOOMD-Blue, simulations can run up to 200 times faster on a GPU enabled computer as compared to a standard desktop computer. By combining the two speed gains from coarse graining the simulation and using GPUs for the calculations, the simulation executes at speeds at least 40,000 times faster than conventional atomistic simulations and approaches the speed of other more commonly used simulation techniques for BCPs. The simulation has shown comparable results to well-known mean field studies. For ideal symmetric copolymers, our model reproduces mean-field order-disorder transition curves. The model also shows similar results to mean-field studies of domain scaling degree of polymerization N in both the strong and weak segregation regimes. Once the MD simulation model was validated, it was used to investigate a series of practical questions related to DSA using block copolymers. This includes: (1) the effect of blending BCPs with other components such as homopolymers, (2) use of tri-block and higher multiplicity block copolymers for DSA, (3) predicting phase diagrams for BCPs where each block has large differences in density or cohesive energy density. Our model will be fully described and highlights of the extensive DSA studies performed using it will be discussed.

(505a) Using Molecular Simulation to Understand Interfacial Phenomena

Errington, J. R., University at Buffalo, The State University of New York

Originally presented on: 11/6/2013 12:30:00 - 13:10:00

Fluids regularly interact with solid substrates in industrial processes, natural phenomena, and throughout our daily experiences. Examples vary from the use of liquid detergents to clean stained fabric to the collection of rain water onto tree leaves. Over the past few decades molecular simulation has emerged as an important tool for understanding the behavior of fluids in the presence of a surface and for predicting interfacial properties. However, progress has stalled, at times, due to an inability to capture relevant free energy information over a wide range of conditions. I will discuss how advanced sampling schemes (e.g. transition matrix Monte Carlo) can be used to overcome this problem, leading to a robust and efficient means for studying interfacial phenomena. Examples of the application of this approach to locate surface phase transitions, examine the effect of nanoscale surface roughness on wetting behavior, and probe the wetting behavior of ionic fluids will be discussed.

(505b) Theory and Simulation Studies Relating Molecular Features to Morphology in Polymer Nanocomposites

Jayaraman, A., University of Colorado, Boulder
Originally presented on: 11/6/2013 13:10:00 - 13:50:00

Polymer nanocomposites are a class of materials that consist of a polymer matrix embedded with nanoscale fillers or additives that enhance the inherent properties of the matrix polymer. Extensive research in the past few decades has established that the spatial arrangement or morphology of the additives in the matrix dictates the macroscopic behavior/properties of the polymer nanocomposite. Therefore to engineer polymer nanocomposites for specific applications with target macroscopic properties (e.g. photovoltaics, photonics, automobile parts) it is important to have design rules that relate molecular features to equilibrium morphology of the composite. This talk will present our integral equation theory and molecular simulation studies focused on understanding how chemical and physical features of the nanoscale additives and polymers affect the equilibrium morphology of the nanocomposite. In the first part of the talk I will present our recent work on composites containing polymer grafted nanoparticles, showing how polydispersity in graft and matrix polymer can be used to stabilize dispersion of the nanoparticles within a polymer matrix. In the second part of the talk I will present our recent work relating conjugated polymer architecture and additive-polymer miscibility to morphologies suitable for organic photovoltaic devices.

(505c) Revealing Reactions in Cellulose Pyrolysis

Mayes, H.,
Beckham, G. T., National Renewable Energy Laboratory
Broadbelt, L. J., Northwestern University
Originally presented on: 11/6/2013 13:50:00 - 14:10:00

Biomass fast pyrolysis is a promising technology for the thermal conversion of biomass to liquid fuels. With less-severe operating temperatures than gasification, fast pyrolysis produces a primarily liquid product, which requires less-costly equipment for processing and storage than a vapor product. Currently, improvements to make this technology more economically attractive are hampered by a lack of understanding of the reaction chemistry. Computational molecular science offers tools to examine fleeting transition states, knowledge of which can aid in rational catalyst and process design. Computation also aids in elucidation of individual mechanisms from a large slate of reactions. In the absence of a mechanistic understanding, pyrolysis models rely on lumped components or a fixed-percent conversion and accurately predict very few individual products. Their validity is limited to feeds and operating conditions very similar to those used to provide the experimental training data for the model. Without product speciation, they have limited usefulness for process optimization to improve yields. This talk will share our recent efforts in revealing cellulose pyrolysis reactions using density functional theory. Specifically, we evaluated several competing theories for the formation of the main product of pure cellulose pyrolysis, levoglucosan, and uncovered a new mechanism that yields results consistent with experiment. We also evaluated several key decomposition reactions of glucose, the cellulose monomer and postulated key intermediate, and how they are perturbed by alkali metal.

(505d) The Search for High-Pressure Ice Phases in Nanoconfined Water Via Molecular Simulation

Addington, C. K., University of Toledo
Phillips, K. A., North Carolina State University
Gubbins, K. E., North Carolina State University
Originally presented on: 11/6/2013 14:10:00 - 14:30:00

Water is a curious substance. It has many useful and interesting properties which arise out of the high polarity of the water molecule. (J. Israelachvili and H. Wennerström, *Nature*, 1996, **379**(6562), 219–225.) This polarity leads to a rich phase behavior, with fifteen different crystal structures of ice known. These ice phases display very different properties due to differences in density, proton ordering, and hydrogen bond networks. For example, the fact that the dielectric constant of ice varies widely between phases has been known for some time (G. Wilson, R. Chan, D. Davidson, and E. Whalley, *The Journal of Chemical Physics*, 1965, **43**, 2384.) and is a result of the strong proton ordering of some ice phases. Water, like other substances, displays interesting properties under confinement. Confinement of water on the molecular level has been experimentally shown to affect phase behavior. (M. Jazdzewska et al., *Physical Chemistry Chemical Physics*, 2011, **13**(19), 9008–9013.) In addition to confinement, the presence of an electric field changes the phase diagram of water because of enhancement or depletion of the hydrogen bonding network. (J. Aragones et al., *Physical Review Letters*, 2011, **107**(15), 155702.) Understanding the phase behavior of water in charged confinement is of interest to several fields, namely biological ion channels and energy storage devices. We have studied the nucleation of ice phases and the structure of water in charged porous carbons to better understand the properties of water in confinement, which are paramount in the effective design of these materials.

(505e) Some Surprises Arising From Atomistic Simulation of Protein Dynamics On the Millisecond Timescale: Protein Folding and Conformational Change of Kinases and GPCRs

Pande, V., Stanford University

Originally presented on: 11/6/2013 14:30:00 - 14:50:00

A key challenge in molecular simulation is reaching experimentally relevant timescales with models that are sufficiently detailed to quantitatively predict experiment. Indeed, most atomistic simulations typically only reach the nanosecond to microsecond timescale, whereas experimentally relevant timescales for many phenomena are on the millisecond to second timescale.

We describe a new approach for simulating long timescale dynamics using Markov State Models (MSMs). MSMs are a systematic sampling scheme to both get to long timescales as well as to gain insight from the results. Indeed, this approach allows a cluster of GPUs to simulate a millisecond of aggregate dynamics and to use that data to quantitatively predict experiment, indicate the degree of robustness of results, and yield novel insights.

Finally, I will demonstrate this method with applications to all-atom molecular simulations on the millisecond timescale and beyond, with applications to protein folding, protein misfolding in Alzheimer's Disease, and protein conformational change in disease-relevant drug targets of GPCRs and kinases.

(556a) Invited Talk: Molecular Simulation of "Real World" Polymer Systems
Greenfield, M. L., University of Rhode Island
Originally presented on: 11/6/2013 15:15:00 - 15:45:00

Molecular simulations intrinsically focus on idealized systems in which the size and chemistry of every molecule is well defined. "Real world" polymers differ from such "textbook" polymers through the presence of additives, fillers, flame retardants, chemically distinct end groups, and so on. Yet how can such effects be incorporated into a molecular simulation framework? Our group has conducted simulations that attempt to quantify aspects of these real world effects within well-defined molecular models. Recent rotational isomeric state calculations have focused on elastomer configurations and on how the population of conformers changes as a function of shape. We are currently extending this to quantify effects of fillers. In molecular dynamics simulations, we have quantified how a small molecule additive can affect local chain dynamics, altering relaxation rates in its immediate environment. This was quantified as a function of vector distance from the additive. Approaches towards modeling flame retardants present at high concentrations may also be described.

(556b) Solvent Diffusion and Drying Properties of Amorphous Polymers Via Molecular Simulation

Xi, L., Massachusetts Institute of Technology
Trout, B. L., Massachusetts Institute of Technology
Originally presented on: 11/6/2013 15:45:00 - 16:05:00

Continuous manufacturing of polymer films requires the efficient removal of solvent contents: enhanced drying performance is one of the major considerations in selecting the film formulation. The total drying time is often dominated by the final stage of drying where a small amount of moisture needs to diffuse through a film of mostly solid components. In this presentation we will first discuss the fundamental mechanism of solvent diffusion at this concentration regime. Diffusion of small molecules in amorphous polymers is known to follow a form of so-called hopping motion: penetrant molecules are trapped in microscopic cavities for extended time periods, and diffusion is made possible by rare but fast jumps between neighboring cavities. We study the molecular mechanism of these hopping motions by statistically analyzing the transitional path ensembles associated with individual hopping events. Our investigation shows that a hopping event starts when the solvent molecule is able to (1) find a pathway between cavities that it is able to push open and (2) break the hydrogen bonds it has formed with the polymer matrix in the original cavity. We then quantify the structural features of polymer matrices that are important to the solvent diffusion according to this mechanism. These quantities are evaluated for different polymer formulations, and their capability of predicting the drying properties is tested through comparison with experimental drying curves. These investigations will provide guidelines for the formulation design of polymer film casting.

(556c) Simulations of Bond Rupture in Normal Alkanes Under Tensile Molecular Deformation Using Reactive Potentials

Nouranian, S., Mississippi State University
Gwaltney, S. R., Mississippi State University
Tschopp, M. A., Army Research Laboratory
Baskes, M. I., University of California, San Diego
Horstemeyer, M. F., Mississippi State University
Originally presented on: 11/6/2013 16:05:00 - 16:25:00

The molecular modeling of failure and fracture phenomena, such as crazing and crack formation, in polymers is of key interest in the multiscale modeling of these materials incorporating damage evolution and progression at different scales. In this study, preliminary simulations of bond rupture in single normal alkane molecules under tensile molecular deformation were performed using three reactive potentials, i.e., modified embedded-atom method (MEAM), second-generation reactive empirical bond order (REBO), and reactive force field (ReaxFF). The higher alkanes in the series provide reasonable representation of a polyethylene chain. The MEAM potential, a semi-empirical many-body potential traditionally used for metals, was recently parameterized for saturated hydrocarbons and is now suitable for the atomistic simulations of complex reactive and non-reactive metallic/organic material systems. The MEAM-, REBO-, and ReaxFF-reproduced energetics of molecular deformation and bond rupture as well as the structural breakdown of a homologous series of alkane molecules from normal propane (n-C_3H_8) to normal undecane (n-$C_{11}H_{24}$) were compared to those reproduced by first-principles (FP) methods. The molecules were initially energy-minimized using the Polak-Ribiere conjugate gradient method. For convenience, the coordinates of the first carbon atom was fixed at the origin, while all the atomic coordinates were scaled in the direction of the molecular backbone (the vector between the first and terminal carbon atoms) with a distance increment (ΔL) of 0.1 Å. The molecules were energy-minimized after each affine scaling of the atomic coordinates, while the coordinates of the first and terminal carbon atoms were fixed. The deformation continued for 100 distance increments (ΔL =10 Å). The bond rupture energies and the resulting molecular fragments (free radicals) differed significantly among the three potentials. While both MEAM and ReaxFF predicted a smooth rupturing instance and fragment formation for all the molecules, REBO predicted relatively high strain energy before rupture that would increase with increasing size of the molecules as well as the formation of many smaller molecular fragments, especially for higher alkanes. Furthermore, REBO allowed significantly bent C-C-C bond angles (almost 180°) to manifest before rupture. ReaxFF almost exclusively led to a methyl radical formation as the smaller molecular fragment resulting from bond rupture, while MEAM predicted the formation of an ethyl radical, especially for the rupture of higher alkanes.

To further study the strain distribution in the molecules during molecular deformation, another set of simulations were run, where only the coordinates of the terminal carbon atoms in the molecules were changed as the molecules were strained. At each intermediate step, the first and terminal carbon atom coordinates were again fixed, while the energies of the molecules were minimized. The average C-C bond distances and C-C-C bond angles were calculated for the molecules up to the moment of rupture and compared with FP calculations.

This study provides insight into the behavior of different alkane molecules under tensile deformation as simulated using the currently available reactive potentials with the potential applicability to the fracture in high polymers, such as polyethylene. Advantages and disadvantages of these potentials in capturing the essential energetics and structural details of the bond rupture phenomena are further compared and contrasted. The results of this study further highlight the essential characteristics of bond breaking for the three reactive potentials.

(556d) The Ab Initio Design of Efficient Photoinitiators

Musgrave, C. B., University of Colorado, Boulder
Lim, C. H., University of Colorado Boulder
Gong, T., University of Colorado at Boulder
Holder, A., University of Colorado at Boulder
Bowman, C. N., University of Colorado
Originally presented on: 11/6/2013 16:25:00 - 16:45:00

We have employed ab initio quantum chemical methods to model a set of both existing and new photoinitiators (PI) and synthesized and characterized several PIs that are predicted to be promising. Calculations of existing PIs are used to benchmark the specific computational methods as well as to develop a fundamental understanding of the various phenomena involved in the photoinitiation process. We then use quantum chemistry to computationally prototype a series of new PIs by calculating a set of properties for each case, including; absorption peak, oscillator strength/absorbtivity, bond dissociation energy, radical reactivity and properties related to the rate of intersystem crossing. We have synthesized and characterized several new PIs, including derivatives of the phosphine oxide and the benzoyl germanium PIs. These experiments confirm the ab initio predictions and provide additional feedback to the computational prototyping effort.

(556e) All-Atom Molecular Dynamics Simulations of Poly(N-isopropylacrylamide) Grafted Architectures

Kamath, G., University of Missouri-Columbia
Deshmukh, S., University College Dublin
Mancini, D. C., Argonne National Laboratory
Sankaranarayanan, S., Argonne National Laboratory
Originally presented on: 11/6/2013 16:45:00 - 17:05:00

All-atom Molecular Dynamics Simulations of Poly(N-isopropylacrylamide) Grafted Architectures

Ganesh Kamath[3], Sanket A. Deshmukh[1], Derrick C. Mancini[2], Subramanian K.R.S. Sankaranarayanan[1]

[1]Center for Nanoscale Materials, Argonne National Laboratory, Argonne, IL 60439

[2]Physical Sciences and Engineering, Argonne National Laboratory, Argonne, IL 60439

[3]Department of Chemistry, University of Missouri-Columbia, Columbia 65211

Poly(N-isopropylacrylamide) (PNIPAM) is a thermosensitive polymer that is well-known for its lower critical solution temperature (LCST) around 305K. Below the LCST, PNIPAM is soluble in water, and above this temperature, polymer chains collapse and transform into a globule-state. In this study, we have carried out MD simulations of PNIPAM polymer chains consiting of 60 monomer units grafted on a gold nanoparticle and a planar surface to study the effect of curvature and temperature on the polymer conformations. We have also studied effect of grafting density and size of nanoparticle on the coil-to-globule collapse shown by PNIPAM above the LCST. Studied system consisted of 100 K to ~3 million atoms. All the simulations were carried out below and above the LCST of PNIPAM, namely, at 275 and 325 K. Simulation trajectories were analyzed for structural and dynamical properties such as radius of gyration of PNIPAM chains, hydrogen bond life-times, and residence time probability of water molecules.

Use of the Center for Nanoscale Materials was supported by the U. S. Department of Energy, Office of Science, Office of Basic Energy Sciences, under contract No. DE-AC02-06CH11357.

(556f) Atomistic Molecular Dynamics Simulations of Precise Ionomers

Bolintineanu, D. S., University of Minnesota
Stevens, M. J., Sandia National Laboratories
Frischknecht, A. L., Sandia National Laboratories
Originally presented on: 11/6/2013 17:05:00 - 17:25:00

Ionomers are being investigated as potential solid electrolytes in battery applications, due to their unique electrical properties. However, the relationships between ionomer chemistry, morphology and ion transport are poorly understood, which has hindered the development of ionomer-based batteries. To this end, we report atomistic molecular dynamics (MD) simulations of a series of polyethylene-based model ionomer melts, for which spacing between functional groups is precisely controlled. We vary the polymer-bound ion chemistry (methylimidazolium, sulfonate and carboxylate), the counterion type, the neutralization level, and the length of the spacer. Where experimental data are available, comparison to the structure factor computed from the simulations is highly favorable. The simulations provide new insights into the shape, size and composition of ionic aggregates. In particular, we observe a wide variety of aggregate morphologies, ranging from small spherical aggregates to string-like shapes and large percolated networks. The unexpected morphologies of these ionic aggregates imply the need for a new interpretation of scattering spectra for these materials. As expected, counterion dynamics depend strongly on the nature of the polymer-bound ionic group. Overall, these simulations elucidate several important connections between physicochemical properties of ionomers, aggregate morphologies, and the nature of counterion transport.

(556g) MD Simulation Study of Polyelectrolyte Chain Collapse in Aqueous Mixed Solvent Due to Hydrophobicity: PAA-Water-Ethanol

Sappidi, P., Indian Institute Of Technology Madras
Natarajan, U., Indian Institute Of Technology Madras
Originally presented on: 11/6/2013 17:25:00 - 17:45:00

Molecular dynamics (MD) simulations were performed to investigate conformational transitions, especially swelling and collapse, of poly(acrylic acid) PAA in dilute water-ethanol solution, as a function of ethanol concentration (leading up to phase separation level) and degree-of-ionization (i.e. charge density). Results were compared to available experimental data from literature on this system. With increase in ethanol concentration, chain swelling (increase of radius of gyration) is observed for un-ionized chain (f=0), and on the contrary for partially and fully ionized cases chain collapse is observed. Snapshot of ionized PAA chain with solvent molecules at 90 vol% ethanol at which significant reduction of $\langle R_g \rangle$ occurs is shown in figure 1, and the variation of $\langle R_g \rangle$ with charge density and ethanol concentration is shown in figure 2. The addition of ethanol to the solution results in expansion of the compact-form for un-ionized chain due to hydrophobic effect. The inter and intra-molecular hydrogen bonds were analyzed. With an increase in ethanol concentration the number of hydrogen bonds between un-ionized PAA and ethanol molecules shows increase and that between PAA and water molecules shows a decrease. For the ionized PAA case, chain shrinkage is found to be influenced by extent of intra-chain and intermolecular hydrogen bonding with water as well as ethanol. We observe that the number of intra-chain hydrogen bonds of PAA increases with addition of ethanol to the solution, in the entire range of ionization (charge density $0 < f < 1$). The localization of ethanol molecules near the PAA backbone at higher levels of ethanol is facilitated by a displacement of water molecules indicating presence of specific ethanol hydration shell, confirmed by results of the RDF curves and coordination number calculations. This behavior, controlled by hydrogen bonding provides a significant contribution to such a conformational transition behavior of the polyelectrolyte chain. The canonical electrostatic potential energy contributions were analyzed. The increase in overall ethanol content leads to formation of ethanol clusters near the PAA chain, thereby accompanying chain shrinkage. The interactions between counter-ions and charges on the PAA chain also influence chain collapse, by way of intermolecular attraction due to dipolar interactions leading to formation of multiplets and a decrease in the osmotic pressure of the chain given the decrease in number of mobile counter-ions.

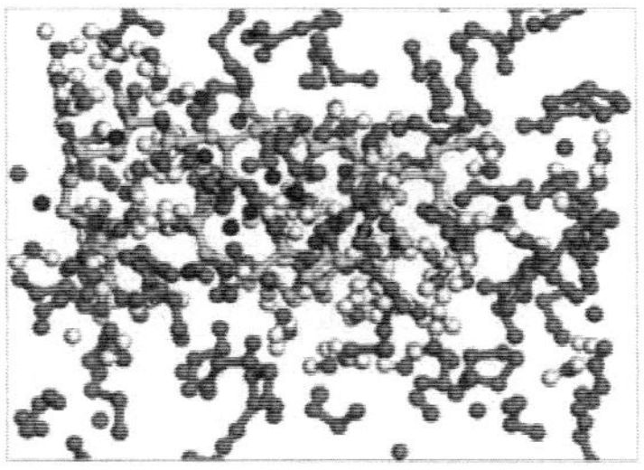

Fig 1. Snapshot of solution containing fully ionized PAA chain (ethanol: water=90:10 by vol.).

(a)

(b)

(c

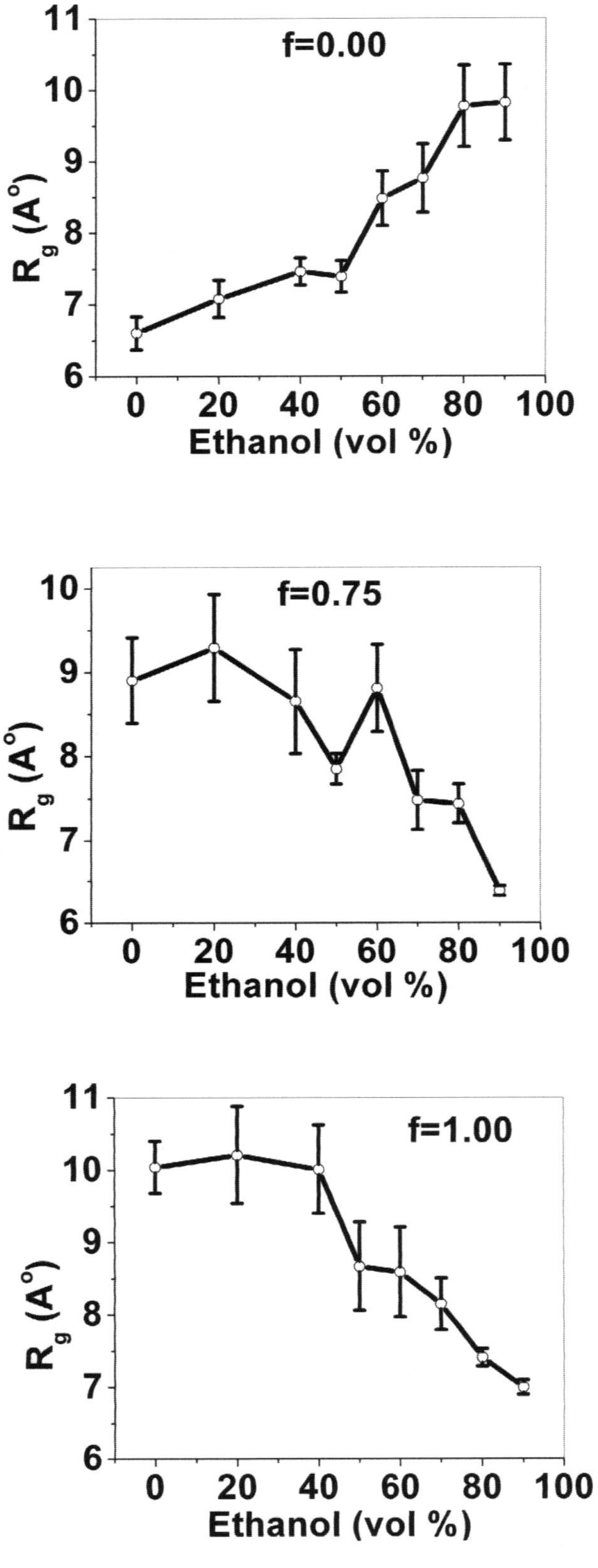

Fig 2. Radius of gyration <R_g> with increase in ethanol fraction for (a) f = 0, (b) f=0.75 and (c) f=1.

(557a) Materials-Scale Implications of Solvent and Temperature On [6,6]-Phenyl-C61-Butyric Acid Methyl Ester (PCBM): A Theoretical Perspective

Tummala, N. R., The University of Oklahoma
Mehraeen, S., Stanford University
Fu, Y. T.,
Risko, C., Georgia Institute of Technology
BrÃ©das, J. L., Georgia Institute of Technology
Originally presented on: 11/6/2013 15:15:00 - 15:35:00

Packing and orientation of pi-conjugated molecules within the active layer of organic solar cells or other organic electronics devices, effectively controls the performance of these devices.[1-7] For this reason, the ability to understand and predict in detail how these molecules pack on their own and in the presence of trace solvents is an important step on the way to increase the overall device efficiency. Here, we focus on [6,6]-phenyl-C_{61}-butyric acid methyl ester (PCBM), an electron-acceptor material of choice in organic solar cells. Through molecular dynamics (MD) simulations, we examine the effects of temperature and trace solvents on the packing and morphological features of bulk PCBM.[8] Solubility (miscibility) parameters, melting and order-disorder transitions, surface energies, and orientational distributions as a function of different starting configurations are discussed. On the basis of the derived morphologies, we combine electronic-structure theory and a kinetic Monte Carlo approach to evaluate the parameters impacting electron mobility in crystalline and amorphous PCBM structures, which are used to explain the range of electron mobilities reported in the literature.

[1] S. P. Tiwari, X.-H. Zhang, J. W. J. Potscavage, B. Kippelen, J. Appl. Phys. 2009, 106, 054504.

[2] V. D. Mihailetchi, J. K. J. van Duren, P. W. M. Blom, J. C. Hummelen, R. A. J. Janssen, J. M. Kroon, M. T. Rispens, W. J. H. Verhees, M. M. Wienk, Adv. Funct. Mater. 2003, 13, 43.

[3] M. T. Rispens, A. Meetsma, R. Rittberger, C. J. Brabec, N. S. Sariciftci, J. C. Hummelen, Chem. Commun. 2003, 2116.

[4] L. Zheng, Y. Han, J. Phys. Chem. B 2012, 116, 1598.

[5] F. Frigerio, M. Casalegno, C. Carbonera, T. Nicolini, S. V. Meille, G. Raos, J. Mater. Chem. 2012, 22, 5434.

[6] L. Chang, H. W. A. Lademann, J.-B. Bonekamp, K. Meerholz, A. J. MoulÃ©, Adv. Funct. Mater. 2011, 21, 1779.

[7] D. L. Cheung, A. Troisi, J. Phys. Chem. C 2010, 114, 20479.

[8] N. R. Tummala, Y. T. Fu, S. Mehraeen, C. Risko, J. L. Bredas, Adv. Funct. Mater. 2013, DOI:10.1002/adfm.201300918.

(557b) Optimizing the Fabrication of All-Organic Solar Cells: A Kinetic Monte Carlo Study of Fullerene Adsorption in Phthalocyanine-Based Covalent Organic Frameworks

Koo, B., Cornell University
Clancy, P., Cornell University
Originally presented on: 11/6/2013 15:35:00 - 15:55:00

Optimizing the Fabrication of All-Organic Solar Cells: A Kinetic Monte Carlo Study of Fullerene Adsorption in Phthalocyanine-based Covalent Organic Frameworks

Brian Koo, Paulette Clancy

Chemical and Biomolecular Engineering, Cornell University

The class of materials known as covalent organic frameworks (COFs)[1] has gained traction recently in the field of photovoltaics. These two-dimensional polymers, originally studied extensively for small gas storage, have high surface area to volume ratios that rival the capacities of metal-organic frameworks and zeolites. Since the incorporation of chromophores into the framework, specifically phthalocyanine groups,[2] optoelectronic applications of COFs became viable. The advancement of synthesizing 2D COFs as thin films on graphene-covered substrates[3] enabled the processing of COFs as a hole-conductor for organic photovoltaic cells. Indeed, recently, a solar cell was made from a combination of PCBM (electron carrier) and a thienothiophene-based COF (hole carrier) but displayed a disappointingly low solar efficiency.[4] Our study explores how to improve this efficiency by studying the filling of thin film phthalocyanine COFs with fullerene molecules (C60) to create prototypical ordered heterojunction solar cells. COFs are a desirable material since they allow, in principle, the ability to transport free charge carriers perpendicular to its stacked layers, and its inherent ability to direct the structural characteristics of semiconducting guest molecules within its pores, such as fullerene.

Many complementary computational studies have probed the nature of stacking of 2D COF layers, given the importance of angstrom-sized offsets on the hole/electron mobility. Adjacent layers adopt a 1.0-1.7 angstrom offset[5] in weakly interacting layers, leading to a stack of 2D layers. This, in turn, induces a large potential energy barrier on the pore surface, which limits molecular dynamics (MD) studies of the adsorption of guest molecules within the pores, since transport within micropores occurs purely by surface diffusion. In a previous study, we predicted that layers are offset with no preferred stacking direction; therefore layers are equally likely to adopt stacking patterns that only have local 3-layer persistence, e.g., helical, zigzag, and staircase stacking, each with locally defined lattice sites. Here, we calculate the rates of diffusion between these lattice sites from the free energy barriers extracted from steered molecular dynamics simulations, and find the location and stability of lattice sites using a minimization procedure. From these data, we use Kinetic Monte Carlo to access far longer timescales of adsorption and diffusion than are possible with MD to find the equilibrated packing structure of fullerene as prototypical n-type guest molecules. We also explore the role of solvent and the effect of pore size on the lattice mismatch between COF and C_{60}. Finally, we predict a theoretical maximum packing density and compare it to experiment[4] to estimate the amount of improvement possible with this arrangement of solar cell component materials.

1. Cote, A. P. *et al.* Porous, crystalline, covalent organic frameworks. *Science* **310**, 1166?70 (2005).

2. Spitler, E. L. & Dichtel, W. R. Lewis acid-catalysed formation of two-dimensional phthalocyanine covalent organic frameworks. *Nat. Chem.* **2**, 672?677 (2010).

3. Colson, J. W. *et al.* Oriented 2D Covalent Organic Framework Thin Films on Single-Layer Graphene. *Science* **332**, 228?231 (2011).

4. Dogru, M. *et al.* A Photoconductive Thienothiophene-Based Covalent Organic Framework Showing Charge Transfer Towards Included Fullerene. *Angew. Chem.* **125**, 2992?2996 (2013).

5. Koo, B. T., Dichtel, W. R. & Clancy, P. A classification scheme for the stacking of two-dimensional boronate ester-linked covalent organic frameworks. *J. Mater. Chem.* **22**, 17460-17469 (2012).

(557c) Insights Into Ionic Liquids Pretreatment On Lignocellulosic Biomass From Molecular Modeling Studies

Parthasarathi, R., Sandia National Laboratories
Shi, J., Sandia National Laboratories
Sun, N., Joint BioEnergy Institute
Sale, K., Joint BioEnergy Institute, Emeryville, CA 94608 and Sandia National laboratories, Livermore, CA
Simmons, B. A., Joint BioEnergy Institute, Emeryville, CA 94608 and Sandia National laboratories, Livermore, CA
Singh, S., Joint BioEnergy Institute, Emeryville, CA 94608 and Sandia National laboratories, Livermore, CA
Originally presented on: 11/6/2013 15:55:00 - 16:15:00

A deeper mechanistic understanding of the lignocellulosic biomass dissolution in ionic liquids (ILs) is desired for the development of biorefining applications to enhance the efficiency of biofuel production. Previous work at the Joint BioEnergy Institute (JBEI) has demonstrated that certain ionic liquids (ILs) are efficient solvents for pretreating many types of biomass with high sugar yields. As a part of a broader effort to develop novel ILs and to improve IL process selectivity, biocompatibility, and economics, here we use an integrated experimental and computational approach to gain a fundamental understanding of the chemistry of biomass deconstruction. We use a multi-resolution computational approach to analyze (A) ions, and ion pairs (i.e. ILs) interactions with model lignin compound (B) interactions between ILs with crystalline cellulose, and (C) developing predictive models for ILs reactivity. In this study, quantum chemical calculations were carried out to probe the interactions of four ILs composed of 1-ethyl-3-methylimidazolium ($[C_2mim]^+$) and cholinium ($[Ch]^+$) cations combined with acetate ($[OAc]^-$) and lysinate ($[Lys]^-$) anions with dilignol. The mechanism for explaining the dissolution of cellulose in different $[C_2mim][OAc]$:water loading is deduced from the comprehensive fully-atomistic molecular dynamics simulations. Finally, we examined the relationship between the predicted Kamlet and Taft (K-T) parameters (dipolarity/polarizability (π*), hydrogen bond donor acidity (α), and hydrogen bond basicity (β)) and experimentally tested K-T parameters on a series of imidazolium based and amino acid based ILs. The models developed in these studies will guide the targeted design of ILs on solubilizing lignocellulosic biomass and will advance the development of IL pretreatment technologies.

(557d) Deriving Coarse-Grained Models for Crystalline Cellulose Based On Force Matching

Markutsya, S., Iowa State University
Devarajan, A., Iowa State University
Gordon, M., Iowa State University
Windus, T. L., Iowa State University
Lamm, M. H., Iowa State University
Originally presented on: 11/6/2013 16:15:00 - 16:35:00

A fundamental understanding of the intermolecular forces that bind polysaccharide chains together in cellulose is crucial for designing efficient methods to overcome the recalcitrance of lignocellulosic biomass to hydrolysis. Because the characteristic time and length scales for the degradation of cellulose by enzymatic hydrolysis or chemical pretreatment span orders of magnitude, it is important to closely integrate the molecular models used at each scale so that, ultimately, one may switch seamlessly between quantum, atomistic, and coarse-grained descriptions of the system. As a step towards that goal, multiscale coarse-grained models for polysaccharide chains in cellulose microfibers are being developed. In this work, the force matching method is used to derive effective coarse-grained forces from all-atom molecular dynamics trajectories. The force matching method systematically optimizes coarse-grained forces from all-atom simulation trajectories without fitting them to a predetermined analytical function. Several coarse-grained mapping schemes were evaluated and tested against the same all-atom reference system. The properties that were compared include crystalline lattice parameters, radial distribution functions, and root mean square fluctuations and displacements. The success of the coarse-grained molecular dynamics simulation in predicting the properties computed for the all-atom crystalline cellulose reference system was dependent on the choice of coarse-grained mapping scheme. This dependence on mapping scheme has been observed before for glucose solutions (Markutsya, *et al.*, 2012) and it is even more pronounced for this application due to the crystalline structure. To overcome this challenge, a rationale strategy for designing coarse-grained mapping schemes in macromolecular crystals is developed and presented. Using this approach, successful agreement is observed between the coarse-grained and all-atom molecular dynamics simulations for crystalline cellulose.

S. Markutsya, Y. Kholod, A. Devarajan, T. L. Windus, M. S. Gordon, and M. H. Lamm, Theor. Chem. Accounts 131, 1162 (2012).

This research is sponsored by the U.S Department of Energy (USDOE) Scientific Discovery through Advanced Computing (SciDAC) program through USDOE's Office of Advanced Scientific Computing Research (ASCR) and Biological and Environmental Research (BER), and performed at the Ames Laboratory, FWP AL-08-330-039.

(557e) Breaking Down Cellulose â€“ the Catalytic Mechanism of a Key Cellulase Elucidated By Path Sampling

Knott, B. C., National Renewable Energy Laboratory
Crowley, M. F., National Renewable Energy Laboratory
Beckham, G. T., National Renewable Energy Laboratory
Originally presented on: 11/6/2013 16:35:00 - 16:55:00

The enormous reserves of carbohydrates found in plant cell walls represent a promising potential source of renewable energy. Deconstruction of these molecules, either thermochemically or biochemically, precedes the ability to transform them to commodity chemicals or fuels. The biochemical route involves the utilization of enzyme cocktails that synergistically break down the biopolymers. Glycoside hydrolases (GHs) are responsible for myriad transformations of carbohydrate glycosidic linkages including polysaccharide turnover in the biosphere. GHs typically hydrolyze glycosidic bonds via mechanisms that either invert or retain the stereochemistry of the anomeric carbon. Retaining mechanisms consist of two elementary steps that include the formation of an enzyme-glycosyl intermediate and a deglycosylation step wherein a water molecule attacks the anomeric carbon to complete the catalytic cycle. Here we examine the two-step retaining mechanism in the *Hypocrea jecorina* GH Family 7 cellobiohydrolase (Cel7A), a key cellulose-degrading enzyme. Cel7A is often a major component in industrial cellulase cocktails and the target of cellulase engineering studies. We utilize transition path sampling with mixed quantum-mechanics/molecular mechanics (QM/MM) simulations to harvest many trajectories for the glycosylation and deglycosylation steps. Subsequent analysis of these trajectories via Likelihood Maximization yields the reaction coordinate for each step, which is verified using the histogram test. The reaction coordinates involve forming/breaking bonds, as well as less-intuitive elements such as catalytic residue orientation and possible product (i.e. cellobiose) assistance. From there, we compute the full free energy surface for the two-step mechanism, including the movement of the cellobiose product from the product of the glycosylation step mode to the conformation that occurs in between the two catalytic reactions. Additionally, we quantify the puckering of the -1 glucosyl residue along the reaction pathway. Overall, this study provides direct insight into the full catalytic mechanism for a two-step, retaining cellobiohydrolase.

(557f) Cellulase Peripheral Loops Facilitate Complexation On Cellulose Via Whole-Path Stabilization

Lin, Y.,
Chu, J. W., University of California, Berkeley
Crowley, M. F., National Renewable Energy Laboratory
Beckham, G. T., National Renewable Energy Laboratory
Himmel, M. E., National Renewable Energy Laboratory
Originally presented on: 11/6/2013 16:55:00 - 17:15:00

The structure-function relationships of complexation of cellodextrin chains from cellulose crystals into the active site of a glycoside hydrolase family 7 endoglucanase catalytic domain (Cel7B CD) are revealed via atomistic molecular dynamics simulations, reaction path optimization, and free-energy calculations. In addition, we characterized the evolving enzyme-microfibril interactions through the complexation process by analyzing the atomistic simulations with a coarse graining procedure. We found that the semi-open active site of Cel7B CD exhibits similar barriers and free energies of complexation over two distinct pathways, namely scooping of a chain into the active site cleft and threading from the chain end into the channel. Both the scooping and threading paths were observed to proceed with similar energetics in complexing glucan chains of different arrangements on the cellulose surface. An emergent principle is that Cel7B CD facilitates the cellulose processing via whole-path stabilization. The peripheral peptide segments outside the active site cleft are shown to couple to the evolving surface structures of cellulose over the simulated paths. The sequence patterns of this functionality were revealed via the approach of mapping atomistic fluctuations onto the parameters of a coarse-grain model. The adopted strategy of linking physics-based molecular interactions to protein sequence could also be employed to elucidate the principles of other protein machines in processing biopolymers.

(557g) Isomerization and Epimerization of Sugars Using Homogeneous and Heterogeneous Catalysts in Aqueous Media: A Multiscale Molecular Modeling Investigation

Mushrif, S. H., Nanyang Technological University
Krishnam, C. B.,
Originally presented on: 11/6/2013 17:15:00 - 17:35:00

Isomerization and epimerization reactions are of great relevance to the conversion of sugars, as biomass components, to fuels and chemicals and to the production of rare sugars. The economics of the conversion of biomass–derived sugars into furanic intermediates and that of the production of rare sugars based sweeteners and pharmaceuticals depend heavily on sugars transformation. Enzymatic transformation of sugars is expensive and there is a pressing need to develop novel inorganic catalysts that can selectively isomerize/epimerize sugars in a more economical fashion. Recent developments in this field have revealed several potential homogeneous and heterogeneous catalysts that can selectively transform sugars in aqueous media.[1-3] Physico–chemical interactions amongst the sugar, the catalyst and the solvent govern the catalytic mechanism of sugars transformation reactions. In the present work, we investigate the isomerization and epimerization of a model sugar molecule, glucose, using different homogeneous and heterogeneous catalysts in aqueous media. Calculations are performed in condensed phase, adopting the Car–Parrinello molecular dynamics scheme with metadynamics acceleration.[4] Speciation and coordination of different catalysts with glucose, in the presence of solvent, and the corresponding reaction mechanisms and energetics are compared. Solvent dynamics are taken into account while simulating reaction mechanisms and its effect on rate limiting steps of the reactions is studied in detail. Molecular details obtained from the present work will serve as an input for developing and designing novel catalytic materials for sugars transformations in aqueous media, thus providing a bottom-up approach for developing materials for sugars chemistry.

1. Choudhary et al., *J. Amer. Chem. Soc.,* 135, 2013, 3997.
2. Moliner et al., *Proc. Natl. Acad. Sci*, 107, 2010, 6164.
3. Gunther et al., *Nat. Commun*., 3, 2012, 1109.
4. Laio and Gervasio, *Rep. Prog. Phys.*, 71, 2008, 126601.

(591a) Massively Parallel Monte Carlo for Many-Particle Simulations On GPUs

Anderson, J. A., University of Michigan
Jankowski, E., University of Colorado at Boulder
Grubb, T., University of Michigan
Engel, M., University of Michigan
Glotzer, S. C., University of Michigan
Originally presented on: 11/7/2013 8:30:00 - 8:55:00

Current trends in parallel processors call for the design of efficient massively parallel algorithms for scientific computing. Parallel algorithms for Monte Carlo simulations of thermodynamic ensembles of particles have received little attention because of the inherent serial nature of the statistical sampling. In this talk, we present a massively parallel method that obeys detailed balance and implement it for a system of hard disks on the GPU. We reproduce results of serial high-precision Monte Carlo runs to verify the method. This is a good test case because the hard disk equation of state over the range where the liquid transforms into the solid is particularly sensitive to small deviations away from the balance conditions. On a Tesla K20, our GPU implementation executes over one billion trial moves per second, which is 148 times faster than on a single Intel Xeon E5540 CPU core, enables 27 times better performance per dollar, and cuts energy usage by a factor of 13. With this improved performance we are able to calculate the equation of state for systems of up to one million hard disks. These large system sizes are required in order to probe the nature of the melting transition, which has been debated for the last forty years.

(591b) Wefold: A Collaborative Protein Structure Prediction Experiment

Khoury, G. A., Pennsylvania State University-University Park
Liwo, A., Cornell University
Khatib, F.,
Zhou, H.,
Chopra, G.,
Bacardit, J.,
Bortot, L. O.,
Faccioli, R. A.,
Deng, X.,
He, Y.,
Krupa, P., Cornell University
Li, J.,
Mozolewska, M., University of Gdansk
Sieradzan, A.,
Smadbeck, J., Princeton University
Wirecki, T., University of Gdansk
Cooper, S.,
Flatten, J.,
Xu, K.,
Baker, D., University of Washington
Cheng, J.,
Delbem, A. C. B.,
Floudas, C. A., Princeton University
Keasar, C., Ben-Gurion University
Levitt, M.,
Popovic, Z.,
Scheraga, H. A., Cornell University
Skolnick, J.,
Crivelli, S. N.,
Originally presented on: 11/7/2013 8:55:00 - 9:20:00

The protein structure prediction problem continues to elude scientists. Even though many new methods have been introduced, certain classes of prediction targets such as free modeling targets remain a challenge based on blind predictions in the several previous Critical Assessment of Techniques for Protein Structure Prediction (CASP) experiments [1]. To meet this challenge, a large-scale collaborative effort called WeFold was undertaken by thirteen labs, each with their own specialties and approaches in addressing the problem.

In this talk, we will present the different methods or branches collaboratively designed and tested during the WeFold experiment, as well as their predictive ability, outcomes, and lessons learned. Independent branches involved in the collaborative effort yielded several high-ranking predictions among all group and method submissions in CASP10 for human, free modeling (template free), and refinement targets. Remarkably, two WeFold methods were able to produce the very best predictions in the refinement category in two different targets from a starting model for several different accuracy metrics. Contributions were made by junior and seasoned scientists alike in an open and accessible collaborative environment. The contributed methods to WeFold include ICOS [3] for contact prediction (Bacardit Lab); CONCORD [4] for secondary structure prediction, BeST for *ab initio* beta-sheet topology prediction [5], contact prediction [6, 7], ICON for traveling-salesman problem based clustering (Floudas Lab); the online multiplayer game Foldit [8] and Rosetta [9] for sampling and selection (Baker Lab); UNRES [10-14] for sampling via Multiplexed Replica Exchange Molecular Dynamics (Liwo Lab and Scheraga Lab); GOAP [15] for knowledge-based scoring, TASSER [16] for sampling, SPICKER [17] for clustering (Skolnick Lab); KoBaMIN [18, 19] for refinement (Levitt Lab); APOLLO [20] for consensus-based quality assessment (Cheng Lab); Replica-Exchange Molecular Dynamics in GROMACS [21] for sampling (Univ. Sao Paulo, Brazil); and MESHI [22] for quality assessment (Keasar Lab).

Performance of the synergistic branches will be compared with the performance of the base methods that comprise them. In total, the collaboration used over 1.5 million CPU hours and processed and evaluated over 8 million candidate structure models (~100x the size of the protein data bank [23]) in a 3.5 month period. All discussions and data generated during the collaboration are publically accessible at http://www.wefold.org.

References:

1. Dill, K.A. and J.L. MacCallum, *The Protein-Folding Problem, 50 Years On.* Science, 2012. **338**(6110): p. 1042-1046.

2. Soding, J., A. Biegert, and A.N. Lupas, *The HHpred interactive server for protein homology detection and structure prediction.* Nucleic Acids Research, 2005. **33**(Web Server issue): p. W244-8.

3. Bacardit, J., et al., *Contact map prediction using a large-scale ensemble of rule sets and the fusion of multiple predicted structural features.* Bioinformatics, 2012.

4. Wei, Y., J. Thompson, and C.A. Floudas, *CONCORD: a consensus method for protein secondary structure prediction via mixed integer linear optimization.* Proceedings of the Royal Society A: Mathematical, Physical and Engineering Science, 2012. **468**(2139): p. 831-850.

5. Subramani, A. and C.A. Floudas, *β-sheet Topology Prediction with High Precision and Recall for β and Mixed α/β Proteins.* PLoS ONE, 2012. **7**(3): p. e32461.

6. Rajgaria, R., S.R. McAllister, and C.A. Floudas, *Towards accurate residue–residue hydrophobic contact prediction for α helical proteins via integer linear optimization.* Proteins: Structure, Function, and Bioinformatics, 2009. **74**(4): p. 929-947.

7. Rajgaria, R., Y. Wei, and C.A. Floudas, *Contact prediction for beta and alpha-beta proteins using integer linear optimization and its impact on the first principles 3D structure prediction method ASTRO-FOLD.* Proteins, 2010. **78**(8): p. 1825-46.

8. Cooper, S., et al., *Predicting protein structures with a multiplayer online game.* Nature, 2010. **466**(7307): p. 756-760.

9. Leaver-Fay, A., et al., *ROSETTA3: an object-oriented software suite for the simulation and design of*

macromolecules. Methods in enzymology, 2011. **487**: p. 545-74.

10. Czaplewski, C., et al., *Application of Multiplexed Replica Exchange Molecular Dynamics to the UNRES Force Field: Tests with α and α+β Proteins.* Journal of Chemical Theory and Computation, 2009. **5**(3): p. 627-640.

11. He, Y., et al., *Exploring the parameter space of the coarse-grained UNRES force field by random search: Selecting a transferable medium-resolution force field.* Journal of Computational Chemistry, 2009. **30**(13): p. 2127-2135.

12. Liwo, A., et al., *Simulation of protein structure and dynamics with the coarse-grained UNRES force field.* Coarse-Graining of Condensed Phase and Biomolecular Systems, 2008. **1**: p. 1391-1411.

13. Liwo, A., et al., *Modification and Optimization of the United-Residue (UNRES) Potential Energy Function for Canonical Simulations. I. Temperature Dependence of the Effective Energy Function and Tests of the Optimization Method with Single Training Proteins.* The Journal of Physical Chemistry B, 2006. **111**(1): p. 260-285.

14. Liwo, A., et al., *Implementation of Molecular Dynamics and Its Extensions with the Coarse-Grained UNRES Force Field on Massively Parallel Systems: Toward Millisecond-Scale Simulations of Protein Structure, Dynamics, and Thermodynamics.* Journal of Chemical Theory and Computation, 2010. **6**(3): p. 890-909.

15. Zhou, H. and J. Skolnick, *GOAP: A Generalized Orientation-Dependent, All-Atom Statistical Potential for Protein Structure Prediction.* Biophysical Journal, 2011. **101**(8): p. 2043-2052.

16. Zhang, Y. and J. Skolnick, *Automated structure prediction of weakly homologous proteins on a genomic scale.* Proceedings of the National Academy of Sciences of the United States of America, 2004. **101**(20): p. 7594-7599.

17. Zhang, Y. and J. Skolnick, *SPICKER: A clustering approach to identify near-native protein folds.* Journal of Computational Chemistry, 2004. **25**(6): p. 865-871.

18. Rodrigues, J.P.G.L.M., M. Levitt, and G. Chopra, *KoBaMIN: a knowledge-based minimization web server for protein structure refinement.* Nucleic Acids Research, 2012. **40**(W1): p. W323-W328.

19. Chopra, G., N. Kalisman, and M. Levitt, *Consistent refinement of submitted models at CASP using a knowledge-based potential.* Proteins: Structure, Function, and Bioinformatics, 2010. **78**(12): p. 2668-2678.

20. Wang, Z., J. Eickholt, and J. Cheng, *APOLLO: A Quality Assessment Service for Single and Multiple Protein Models.* Bioinformatics, 2011. **27**: p. 1715-1716.

21. Hess, B., et al., *GROMACS 4: Algorithms for Highly Efficient, Load-Balanced, and Scalable Molecular Simulation.* Journal of Chemical Theory and Computation, 2008. **4**(3): p. 435-447.

22. Kalisman, N., et al., *MESHI: a new library of Java classes for molecular modeling.* Bioinformatics, 2005. **21**(20): p. 3931-3932.

23. Berman, H.M., et al., *The Protein Data Bank.* Nucleic Acids Res, 2000. **28**(1): p. 235-42.

(591c) Early Experiences With the Intel Xeon Phi and MIC Architecture

Willmore, F. T., University of Texas at Austin

Originally presented on: 11/7/2013 9:20:00 - 9:45:00

Recent advances in High Performance Computing Architecture include General Purpose Graphics Processing Units, originally co-opted from gaming, have already seen wide adoption as accelerators in computational chemistry and scientific computing. Mindful of the ever-increasing power requirements of HPC as technology pushes towards the exascale, and with the release of the ARM64 standard, curious eyes are on the confluence of companies developing ARM-based processors for battery-powered mobile devices, hoping to leverage another consumer product for scientific computing. Meanwhile, Intel has released its own accelerator solution in the form of MIC (Many Integrated Core) technology as seen in the Intel Xeon Phi product line, now deployed and available to the open science community as TACC Stampede.

As part of its mission of enabling scientific discovery through advanced computation and in addition to the deployment of world-class computing systems, the Texas Advanced Computing Center, a research unit of the University of Texas at Austin, strives to evaluate new hardware architectures for potential applications to the computational science community. Some early results on performance, scaling and discussion of competing new technologies for molecular science applications will be presented.

(591d) Advancing Molecular Dynamics Models for Future Computers: A Case Study of Truncated Electrostatics

Nguyen, T. D., Oak Ridge National Laboratory
Carillo, J. M. Y.,
Dobrynin, A., University of Connecticut
Brown, W. M., Sandia National Laboratories
Originally presented on: 11/7/2013 9:45:00 - 10:10:00

The growing use of GPUs and other coprocessors along with the trend towards heterogeneous computing have made it increasingly desirable to reevaluate computational methods and models with an eye toward approaches that allow for increased concurrency and data locality to fully exploit the potential performance of new computer hardware. Enhanced truncation approaches have been proposed as an alternative method to traditional approaches for evaluating long-range Coulombic interactions, as to improve computational efficiency while allowing for a tunable degree of accuracy. While these methods are particularly well suited for many-core architectures and graphics processing units (GPUs) due to the inherent fine-grain parallelism, care must be taken, particularly for systems with inhomogeneous charge distribution. We compare efficient truncation-based approximations to evaluation of electrostatic forces with the more traditional particle-particle particle-mesh (P3M) method for the molecular dynamics simulation of polyelectrolyte brush layers. We show that with the use of GPU accelerators, large-scale parallel simulations on the hybrid Titan supercomputer using P3M can be 3 times faster due to a reduction in the mesh-size required. Alternatively, using a truncation-based scheme can be up to 4 times faster than GPU-accelerated P3M and results in accurate calculation of shear velocities and disjoining pressures for brush layers. For configurations with highly nonuniform charge distributions, however, we find that it is more efficient to use P3M; for these systems, computationally efficient parametrizations of the truncation-based approach do not produce accurate counterion density profiles or brush morphologies.

(591e) Calculation of Virial Coefficients On Graphics Processors

Schultz, A. J., University at Buffalo, The State University of New York
Barlow, N. S., University at Buffalo, The State University of New York
Feng, C., University at Buffalo
Chaudhary, V., University at Buffalo
Kofke, D. A., University at Buffalo, The State University of New York
Originally presented on: 11/7/2013 10:10:00 - 10:35:00

Evaluation of virial coefficients from a molecular model can be performed by Monte Carlo sampling of configurations of a small number of molecules, typically on the order of ten or fewer. Such calculations can have relatively low memory requirements, and are perfectly parallelizable, making this ideal for implementation on graphical processors and other multicore architectures. We describe our experiences performing such calculations in a variety of contexts. We consider first the use of Mayer Sampling Monte Carlo for computation of coefficients up to at least eighth order (B8) for the Lennard-Jones model, and sixth order (B6) for a semiclassical model of helium. We then examine implementation of a remarkable algorithm due to Wheatley, which is particularly effective in application to hard potentials. We examine the strengths and limitations of Wheatley's algorithm in the context of GPU architectures, and consider its performance for a variety of potential models. Ultimately memory limits the order to which coefficients can be computed; we report on the level of capability provided by existing architectures, and consider prospects for these high-order virial calculations as the hardware continues to improve.

(605a) Reconfigurable Assemblies With Shape Shifting Particles

Nguyen, T. D., Oak Ridge National Laboratory
Glotzer, S. C., University of Michigan
Originally presented on: 11/7/2013 8:30:00 - 8:47:00

Utilizing nanoscale building blocks that can change shapes in response to environmental changes proves to be a promising approach to engineering bio-inspired nanomaterials[1,2]. We propose a generic scheme for shifting particle shapes to induce the assembly and reconfigurability of various ordered structures in two- and three-dimensional spaces. Starting from ordered structures made of spherical and rod-like particles, we alter the particle shape at different rates into other shapes such as square, triangle or zigzag (in 2D)[2] or tetrahedron and octopod (in 3D), and observe the change in the morphology of the structure as a function of the initial and target shapes and shifting rate. Using molecular dynamics simulation, we demonstrate that more efficient assembly, higher packing fractions and ordered structures with fewer defects are among the advantages that shifting the particle shapes can offer. We also discuss the challenges the approach should face compared to the traditional assembly method, i.e. without shape shifting.

References

1. T. D. Nguyen and S. C. Glotzer, ACS Nano, 4, 2585-2595, 2010.

2. T. D. Nguyen, E. Jankowski and S. C. Glotzer, ACS Nano, 5, 8892-8903, 2011.

(605b) Systematic Inference of Self-Assembly Pathways By Nonlinear Machine Learning

Long, A. W., University of Illinois
Ferguson, A. L., University of Illinois
Originally presented on: 11/7/2013 8:47:00 - 9:04:00

Controllable self-assembly has opened a new frontier for the fabrication of sophisticated functional aggregates, such as self-healing membranes for biomedical devices, nanoscale optoelectronics for photonic computing, and "smart" drug delivery vessels. Computer simulations provide a means to probe the structural mechanisms and interactions driving micro and nanoscale self-assembly, but it remains a challenge to identify the dynamic structural rearrangements driving these phenomena. The absence of good order parameters describing assembly has frustrated the establishment of a quantitative theoretical framework explicitly accounting for its thermodynamic and kinetic features, and hindered the rational design of programmable building blocks.

We have confronted this challenge by coupling nonlinear data mining algorithms to molecular simulation trajectories of self-assembly. Our approach enables us to systematically extract kinetically meaningful order parameters for assembly, infer assembly pathways and mechanisms, and quantify fluxes through competing assembly routes. In brief, we map the assembly trajectory to a dynamically updating interaction graph, in which nodes are monomers and edges bonds, enabling us to compute the structural similarity of the partially assembled clusters through graph matching. Using the quality of the match as a proxy measure for the dynamic proximity of pairs of clusters, we use the diffusion map nonlinear machine learning algorithm to extract the underlying self-assembly pathways describing the sequence of structural steps by which the system evolves into its terminal assembled state. The order parameters computed by the diffusion map correspond to the collective modes governing the long time evolution of the system, furnishing a kinetically meaningful low-dimensional subspace in which to infer assembly pathways and construct dynamically meaningful projections of the underlying free energy landscape for assembly.

We have applied our approach to the self-assembly of patchy colloids into icosahedral aggregates as a coarse-grained biomimetic model for viral capsid formation. In good agreement with previous work employing cluster size as a heuristic order parameter, our approach systematically identifies two qualitatively different assembly pathways corresponding to 1) monomeric addition, and 2) budding of aggregates from a liquid-like phase, and we recover kinetically meaningful collective order parameters describing the dynamic structural rearrangements along each path. Further, we quantify the flux through each pathway to compute a quantitative measure of the assembly propensity through each route as a function of temperature and chemistry of the colloidal monomers. Finally, we discuss how "assembly landscapes" may be used to guide the design of building blocks programmed to assemble desired aggregates that are both thermodynamically stable and kinetically accessible, by rational manipulation of the underlying free energy landscape.

(605c) A General Mechanism for Emergent Chirality in Achiral Hard Particle Assembly

Carmichael, S., University of California Santa Barbara
Shell, M. S., University of California Santa Barbara

Originally presented on: 11/7/2013 9:04:00 - 9:21:00

For centuries, chirality has been appreciated as a key component in understanding how matter orders. While intuitively chiral particles can self-assemble into chiral superstructures, it is less clear how achiral particles are driven to do the same. Here we show that there is a general mechanism that explains why many simple achiral particles can assemble into chiral materials. Namely, we use simulations of hard, regular polygons to show that surprisingly subtle particle shape modifications, such as corner rounding, can induce chiral symmetry breaking by deforming the underlying close-packed lattice. Not only does this mechanism quantitatively explain recent experimental results reporting chirality in the hard triangle system, but it offers an intuitive and general way by which chirality might emerge in nature. Moreover, because effective corner rounding can be realized by modulating repulsive interactions, this simple mechanism suggests tremendous potential for creating dynamically tunable chiral surfaces with a variety of applications.

(605d) Quantitative Modeling of DNA Grafted Nanoparticle Self-Assembly

Vo, T.,
Srinivasan, B., Columbia University
Kumar, S., Columbia University
Gang, O.,
Venkatasubramanian, V., Columbia University
Originally presented on: 11/7/2013 9:21:00 - 9:38:00

Nanoparticle super-lattice engineering, which seeks to design DNA-grafted NPs that self-assemble into desired structures, is a growing field of research because of its relevance to optical band gap materials, sensors etc. In this protocol, single-stranded DNA sequences are grafted to NPs. This motif corresponds to a central NP core grafted with chains with “sticky” ends. DNA “sticky” ends across two particles with complementary strands can hybridize into double-stranded DNA, thus creating an emergent network of interconnected NPs, which under certain conditions crystallizes into a given lattice structure. A simple, but popular model for the prediction of DNA-grafted NP crystal structure was devised by Macfarlane. This Complementary Contact Model views DNA-NP complexes as “fuzzy spheres” where contacts between NPs lead to permanent DNA hybridization. However, despite its success, the model is limited in its prediction abilities to four broad classes of structure and cannot differentiate between different crystals within those classes. Here, we propose an extension of the model to allow for grafting of two different types of complementary linkers on each particle and incorporate hybridization between particles for both similar and different particles. These modifications enabled us to predict grafting density of DNA linkers on each particle that will result in a transition from a body-centered cubic structure to a face-centered cubic structure formation. Furthermore, the model can now differentiate between different types of crystal structure from the same class such as a disordered FCC, CuAu, and regular FCC structure – in a manner that is apparently in good agreement with the recent experiments of Gang and his coworkers. These new developments allow for a wider selection of design parameters of DNA-NP mixtures and thus can increase the selectivity of the final structure of the system.

(605e) Fundamental Thermodynamic and Kinetic Models for Self-Assembly of Small Clusters of Colloidal Particles

Thyagarajan, R., Indian Institute of Technology Madras
Sehgal, R. M., University of Massachusetts Amherst
Maroudas, D., University of Massachusetts Amherst
Bevan, M. A., Johns Hopkins University
Ford, D., University of Massachusetts Amherst
Originally presented on: 11/7/2013 9:38:00 - 9:55:00

The self-assembly of finite clusters of colloidal particles into crystalline objects is a topic of technological interest, as a route to produce photonic crystals and other meta-materials. Such assembly problems are also of fundamental scientific interest because they involve thermodynamically small systems, with a number of particles (10 to 1000) that is far below the bulk limit. One assembly method involves the use of thermodynamic variables (e.g. temperature, depletant concentration) or external fields as actuators, to modify the level of interparticle attraction. Robust methods for assembling defect-free target structures will ultimately require reduced-dimension process models that link the particle-level dynamics of the colloids to the actuator states. We have developed a three-part strategy for developing such process models. First, we employ diffusion maps (DMaps) on raw trajectory data to identify slow, low-dimensional manifolds in the system dynamics. Second, we identify convenient observables, or order parameters (OPs), that strongly correlate with the low-dimensional DMap coordinates; this step may involve a feedback loop with the DMap process itself. Third, we use a Fokker-Planck or Smoluchowski formalism to build free energy and diffusivity landscapes in the OPs, which serve as our reduced-dimension process models. We demonstrate our strategy on two colloidal systems that have been synthesized and studied experimentally.

(605f) Patterned Substrates to Direct Self-Assembly of Particle Monolayers

Ferraro, M. E., The University of Texas at Austin
Truskett, T. M., The University of Texas at Austin
Bonnecaze, R. T., The University of Texas at Austin
Originally presented on: 11/7/2013 9:55:00 - 10:12:00

As current lithographic techniques approach practical engineering limits for resolution, directed self-assembly of nanoparticles becomes an attractive scalable nanomanufacturing process for creating ordered arrays of particles at a variety of length scales that could be used both as patterning agents and functional materials. However the roles of interparticle forces and external fields on directed self-assembly of particles is not well understood. In this presentation density functional theory (DFT) and molecular dynamics (MD) simulations are used to explore the use of larger scale patterned substrates to drive smaller scale directed self-assembly of particle monolayers. Square patterned substrates with varying energy barriers at length scales N-fold the final desired particle pitch are considered (N>1). Ranges of N, bulk density and patterned substrate field strength are identified that disrupt the entropically favored hexagonally close-packed lattice and promote square lattice formation for hard-spheres. Event-driven MD simulations are then employed to verify the predictions from DFT and to analyze the dynamics of the self-assembly process. The effects of interparticle forces and other geometries of patterned substrates are considered. These DFT and MD results are used to discuss and define the energetically and kinetically accessible spaces for non-hexagonal lattice formation.

(605g) Molecular Dynamics Simulations of Polymer-Surfactant Interactions

Surampudi, L. N.,
Ashbaugh, H. S., Tulane University
Originally presented on: 11/7/2013 10:12:00 - 10:29:00

The self-assembly of surfactants in the aqueous media is an area of active research. The presence of the polymer molecules gives rise to changes in the solution and the interfacial properties of the surfactant compared to the corresponding polymer-free systems. In this work, we have analyzed the polymer-surfactant aggregation phenomenon by looking at the interactions between the anionic sodium dodecyl sulfate (SDS) surfactant and two non-ionic polymers – poly-(ethylene oxide) (PEO) and poly-(vinylpyrolidone) (PVP) using molecular dynamics (MD) simulations in water medium. From the simulation results, it is observed that the polymer chains adsorb on the hydrophobic regions of the SDS micelle surface in a pattern similar to the beaded necklace model as it was observed in experimental studies. The association of hydrophobic regions in the polymer and micelle suggest that the hydrophobic effect is the driving force for association. As hydrophobic effect is a function of temperature, our research is concentrated on studying the polymer-micelle interactions at various temperatures in the range of 280-360K. In all the temperature studies, the nonionic polymer displayed interaction with the anionic surfactant by adsorbing on the surface region of the micelle. The Surface Accessible Surface Area (SASA) of the micelle adsorbed by polymer (PEO and PVP) was small compared to the surface area of the polymer-free micelle, proving that the polymer shields the micelle from water promoting hydrophobic interactions between the polymer and micelle. The Potential of Mean Force (PMF) between the polymer and the surfactant assembly was also evaluated at various temperatures by pulling short chains of PEO and PVP from the surface of the micelle. The strength of interactions between micelle and polymer per unit surface area of the polymer was analyzed as a function of temperature. These results provide valuable insight into the temperature dependent hydrophobic interactions between nonpolar polymers with the anionic surfactants.

(605h) Nucleation, Growth, and Transformations in DNA-Linked Binary Colloidal Superlattices

Jenkins, I., University of Pennsylvania
Sinno, T., University of Pennsylvania
McGinley, J. T. III, University of Pennsylvania
Crocker, J. C., University of Pennsylvania
Originally presented on: 11/7/2013 10:29:00 - 10:46:00

A promising approach for realizing versatile colloidal assembly is to graft single-stranded DNA oligomer brushes onto the surfaces of particles in order to create tunable attractive interactions between them. Using this concept, micro- and nanoscale particles have now been successfully assembled into bulk crystalline phases, including a variety of ordered, binary superlattice structures. However, there remain open questions regarding the prediction of nucleation and growth behavior in such structures as a function of DNA brush characteristics. Moreover, recent observations of diffusionless transformations in binary colloidal crystallites further motivate a more mechanistic understanding of this class of materials.

Here, we apply a combination of Monte Carlo simulations, Brownian dynamics simulations, and free energy calculations to generate a detailed picture for the assembly of binary colloidal superlattice crystals. We first describe a pseudo-phase diagram that includes both thermodynamic and kinetic influences on homogeneously nucleated colloidal crystallites and demonstrate how kinetic limitations during crystal growth play a direct role in establishing the observed phase. Next, we also present a detailed analysis of diffusionless transformations between body-centered and face-centered cubic configurations of binary colloidal crystals and demonstrate that hydrodynamic factors play an important role in establishing the final structure.

(605i) Role of Symmetry in the Self Assembly of Space-Filling Honeycombs

Schultz, B., University of Michigan, Ann Arbor
Damasceno, P. F., University of Michigan
Engel, M., University of Michigan
Glotzer, S. C., University of Michigan
Originally presented on: 11/7/2013 10:46:00 - 11:03:00

The self-assembly of hard polyhedral particles is driven by the maximization of entropy. As particles are forced into increasingly dense configurations, the system seeks states with maximal entropy, which are often ordered crystalline structures. The notion of directional entropic forces - statistical forces that promote particles to bind or align in certain directions - have been developed to support observations that facets on particles can mimic covalent bonds in atomic crystals. Within this framework, we set out to design particles that would self-assemble into honeycombs, or space-filling arrangements of polyhedra, that are isostructural to atomic crystals. We use the Voronoi tessellation to generate hard, space-filling polyhedra, or "Voronoi particles," that are guaranteed to stabilize the target structure at infinite pressure, analogous to designer potentials that stabilize arbitrary crystals structures with complex radially symmetric potentials. We characterize the thermodynamics, kinetics and self-assembly behavior of these polyhedra and demonstrate how an understanding of the role of symmetry and particle shape can guide design to improve assembly behavior.

(629a) Material Science Meets Rigidity Theory Meets Mechanical Engineering: What Can We Learn About Flexibility of MOFs From the Reduced Topological Representations?

Sarkisov, L., University of Edinburgh
Smit, B., University of Amsterdam
Originally presented on: 11/7/2013 8:30:00 - 8:46:00

Flexibility of Metal-Organic Frameworks (MOFs) has become one of the most fascinating topics in material science and crystal engineering. Indeed, over the years a number of MOF structures have been discovered that exhibit a broad and diverse range of conformational effects, from local rotational freedom of linkers to swelling and negative thermal expansion to large amplitude deformations of the framework, such as breathing effects. Given virtually infinite number of possible MOFs, this behaviour can be exploited to develop smart, stimuli responsive, programmable MOFs for controlled drug release, detection and energy storage. Current development of these applications is hindered by poor understanding of the mechanisms of the deformation processes on molecular level. Here we present a systematic way to detect potential flexibility modes in various classes of MOFs. The approach builds on the reduced topological representations of MOF structures, combined with the rigidity theory analysis. As case studies we consider flexibility modes in IRMOF-1, MIL-53 and several other materials.

(629b) Interrogating Simple Pore Models Using Flat-Histogram Sampling Methods to Understand the Effect of Sorbent Flexibility On Fluid Adsorption

Shen, V. K., National Institute of Standards and Technology
Siderius, D. W., National Institute of Standards and Technology
Originally presented on: 11/7/2013 8:46:00 - 9:02:00

While it is well known that sorbent flexibility can affect observed adsorption behavior, rigid sorbents are still commonly used in the study of adsorption via molecular simulation for both simplicity and practicality. Past attempts at incorporating sorbent flexibility in simulation have largely relied on computationally demanding atomistic descriptions of the sorbent framework. In this work, we present two simple pore models for flexible sorbents. The first model accounts for internal sorbent flexibility in a grand canonical ensemble formulation. The second model accounts for volume fluctuations via an osmotic ensemble formulation. In both cases, we use flat-histogram sampling methods to calculate precisely the relevant multi-dimensional probability distributions, which we will show are directly related to specific thermodynamic potentials and the deformation properties of the sorbent. Using a square-well fluid as the adsorbate, we are able to reproduce experimentally observed adsorption behavior, such as adsorption-induced deformation and breathing phase transitions. More importantly, we are able to identify the physically relevant features (or properties) of the sorbent that lead to these interesting and exploitable sorbent properties.

(629c) First Principles Derived, Transferable Force Fields for CO2 Adsorption in Cation-Exchanged Zeolites

Fang, H., Georgia Institute of Technology
Kamakoti, P., ExxonMobil Research and Engineering
Cundy, S., ExxonMobil Research and Engineering
Li, Q., ExxonMobil Research and Engineering
Ravikovitch, P. I., ExxonMobil Research and Engineering
Paur, C., ExxonMobil Research and Engineering
Aronson, M., Univ. of California, Santa Barbara
Sholl, D. S., Georgia Institute of Technology
Originally presented on: 11/7/2013 9:02:00 - 9:18:00

The development of accurate force fields is vital for predicting adsorption in porous materials. Previously, we introduced a first principles-based transferable force field for CO_2 adsorption in siliceous zeolites (Fang et al. *J. Phys. Chem. C* **2012**, *116*, 10692). In this study, we extend our approach to CO_2 adsorption in cationic zeolites which possess more complex structures. Na-exchanged zeolites are first chosen for demonstrating the approach. These methods account for several structural complexities including Al distribution, cation positions and cation mobility, all of which are important for predicting adsorption. The simulation results are validated with high-resolution experimental measurements of isotherms and microcalorimetric heats of adsorption on well-characterized materials. The choice of first-principles method has a significant influence on the ability of force fields to accurately describe CO_2-zeolite interactions. The PBE-D2 derived force field, which performed well for CO_2 adsorption in siliceous zeolites, does not do so for Na-exchanged zeolites; the PBE-D2 method overestimates CO_2 adsorption energies on multi-cation sites that are common in cationic zeolites with low Si/Al ratios. In contrast, a force field derived from the DFT/CC method performed well. Agreement was obtained between simulation and experiment not only for LTA-4A on which the force field fitting is based, but for other two common adsorbents, NaX and NaY. Recently, we have expanded this approach to other cations including divalent Ca-exchanged zeolites as well as Ca/Na mixed zeolites. The effect of residual H_2O on CO_2 adsorption will also be considered.

(629d) Ab Initio Carbon Capture in Open-Site Metal Organic Frameworks - Force Field Development and Its Applications

Lin, L. C., University of California-Berkeley
Lee, K., Lawrence Berkeley National Laboratory
Kim, J., Lawrence Berkeley National Laboratory
Smit, B., University of Amsterdam
Originally presented on: 11/7/2013 9:18:00 - 9:34:00

Carbon capture and sequestration (CCS) is a viable strategy proposed to mitigate the negative environmental impacts derived from burning fossil fuels. Recently, nano-porous adsorbent materials such as metal-organic frameworks (MOFs) have shown to potentially provide a more energy-efficient way to separate CO_2 out of flue gases. In particular, MOFs that possess unsaturated open metal sites are known to interact strongly with CO_2 molecules, and thus emerging as promising materials for the separation applications. However, from the computational point of view, the strong interaction is poorly modeled by common force fields, resulting in underestimation of the adsorption properties of CO_2 in Mg-MOF-74 by as much as an order of magnitude. Accordingly, a systematic methodology has been proposed to generate accurate force fields using high-level quantum chemical calculations[1]. In this method, interaction energies of a few selected CO_2 configurations computed by quantum calculations are used to properly map into the force field parameters. In our previous study[1], Møller-Plesset second-order perturbation theory (MP2) with representative small framework fragments was used in the force field development. To facilitate the automation of this methodology as well as resolve the intrinsic discrepancies between cluster and periodic calculations, a methodology based on the periodic density functional theory (DFT) calculations is introduced in this work. The calculated binding geometry and adsorption properties from the DFT-derived force fields show excellent agreements with the ones from DFT calculations and experiments, respectively. Finally, we have also utilized the new DFT-derived force field to study the CO_2 dynamics in Mg-MOF-74 and made connections to our previous study[2].

Reference

[1] A. Dzubak, L.-C. Lin, J. Kim, J.A. Swisher, R. Poloni, S.N. Maximoff, B. Smit, L. Gagliardi, Ab initio carbon capture in open-site metal–organic frameworks, *Nat. Chem.* 2012 (4), 810-816.

[2] L.-C. Lin, J. Kim, X. Kong, E. Scott, T.M. McDonald, J.R. Long, J.A. Reimer, B. Smit, Understanding CO_2 Dynamics in Metal-Organic Frameworks with Open Metal Sites, *Angew. Chem. Int. Ed.* 2013 (52), 4410-4413.

(629e) Virial Coefficient Theory for Adsorption of Gas Mixtures

Yang, J. H., University at Buffalo, The State University of New York
Schultz, A. J., University at Buffalo, The State University of New York
Errington, J. R., University at Buffalo, The State University of New York
Kofke, D. A., University at Buffalo, The State University of New York
Originally presented on: 11/7/2013 9:34:00 - 9:50:00

Reliable prediction of the properties of mixtures in the presence of adsorbents is a key capability for the design of adsorption processes. Molecular-based approaches are needed to include the key physical considerations to perform predictive modeling, and several such approaches are available. These include classical density functional theory, integral equation theories, and molecular simulation. Each of these methods has advantages and disadvantages relating to the complexity of physical features that can included in the treatment, as well as computational cost and suitability for analysis. Among this field of considerations a special place is held by methods based on a virial treatment of the properties. Such methods can incorporate almost any level of detail in the molecular model, including molecular flexibility, quantum effects, multibody interactions, *etc.*, and (unlike molecular simulation) yield an analytic equation for the behavior. On the other hand, the range of application of the treatment is limited to low-density phases, which nevertheless are still of practical interest.

Virial-based methods are well established for bulk fluids, where the approach has classically provided capabilities somewhere between thermodynamic models and molecular simulations. This powerful, systematic method can be applied to inhomogeneous systems as well. Even though the treatment of the general imperfect gas can easily extended to the present problem, the study of gas-surface system is far less well developed. The theory was developed by the 1962 papers by Bellemans, who then formulated a Mayer-cluster theory of an inhomogeneous system and showed a unique way to separate the surface terms proportional to the surface area

In the present work, the virial coefficient theory is applied to the gas-surface system. We report a study of gas mixtures adsorbed on solid surface using inhomogeneous virial coefficients. The relevant surface variables for gas-surface system are directly expressed in terms of interactions between molecules themselves, and these molecules with the surface. The Mayer-sampling Monte Carlo (MSMC) method is applied to calculate the cluster integrals which contain surface contributions. We first report the results of calculation of "surface" virial coefficients of a mixture evaluated by using MSMC. Thermodynamic properties such as pressure, isosteric heat of adsorption are calculated by a virial expansion. Also we examine the selectivity at the limit of zero pressure which is directly related to the intermolecular forces. Comparison with molecular simulation demonstrates the range of applicability of the virial treatment.

(629f) An Improved Classical Density Functional Theory for the Description of Adsorption Thermodynamics

Hlushak, S., Vanderbilt University
McCabe, C., Vanderbilt University
Cummings, P. T., Oak Ridge National Laboratory
Originally presented on: 11/7/2013 9:50:00 - 10:06:00

An improved classical density functional theory [1] (CDFT) based on the accurate bulk thermodynamics predicted by the first-order mean-spherical approximation (FMSA) [2] is applied to study the thermodynamics of adsorption of simple gases into porous materials.

The CDTF theory is implemented in multiple dimensions, which allows studies of various confinement geometries, and makes use of Fast Fourier Transform to integrate the convolution integrals appearing during the calculation of the weighted densities. Contributions from the attractive interactions are taken into account using accurate structural properties [3,4,5] and energy-route thermodynamics obtained using the FMSA theory enhanced with the simplified exponential approximation [6,7,8]. The introduced improvements allow us to quickly and accurately calculate isotherms and heats of adsorption.

[1] S. Hlushak, C. McCabe, P. Cummings. J. Chem. Phys., 137, 104104 (2012).
[2] Y. Tang, and B. C.-Y. Lu, J. Chem. Phys., 99, 9828 (1993).
[3] Y. Tang, J Chem. Phys. 127, 164504 (2007).
[4] S. Hlushak, A. Trokhymchuk, S. Sokolowski, J. Chem. Phys. 130, 234511 (2009).
[5] S. Hlushak, A. Trokhymchuk, S. Sokolowski. J. Chem. Phys. 134, 114101 (2011).
[6] Y. Tang, B. C-Y. Lu. AIChE journal, 43, 2215-2226 (1997).
[7] S. Hlushak, A. Trokhymchuk, Cond. Matt. Phys., 15, 23003 (2012).
[8] S. Hlushak, P. Hlushak, A. Trokhymchuk, J. Chem. Phys., 138, 164107 (2013).

(629g) Computational Screening of Microporous Organic Solids: Development of Design Principles and Analysis of BET Theory

Abbott, L. J., Pennsylvania State University
Colina, C. M., Pennsylvania State University
Originally presented on: 11/7/2013 10:06:00 - 10:22:00

Organic molecules and dendrimers of intrinsic microporous (OMIMs and DIMs) are amorphous, organic microporous materials, which are of interest for a variety of applications, including gas separation and storage. They incorporate rigid and awkward structures to promote inefficient packing, similar to polymers of intrinsic microporosity (PIMs). Molecular simulations have been utilized to screen an array of >30 OMIMs and DIMs with a wide variety of structures in order to guide future synthetic efforts. Analysis of the simulations has revealed the role of three structural design aspects for increasing porosity in these types of materials. First, molecular rigidity has been observed to have a profound effect on pore formation, and simulations of ideal rigid-body OMIMs have resulted in materials with surface areas larger by an order of magnitude. Second, efficient packing and molecular interpenetration is hindered by the inclusion of bulky terminal units, such as *tert*-butyl and adamantyl functional groups. Third, molecular structures that are more three-dimensional, instead of planar, introduce greater "internal molecular free volume" and help promote more inefficient packing to increase porosity.

In addition to providing insight into important design principles, our computational screening has enabled a closer look into the applicability of BET theory for deriving apparent surface areas from adsorption data in these types of microporous materials. Specifically, we have compared geometric surface areas measured from the simulations with BET surface areas calculated from simulated nitrogen adsorption isotherms at 77 K, as is done experimentally. Our results have shown a systematic over-prediction by the BET surface areas, especially for the ultramicroporous materials. We utilize the array of intrinsically microporous materials we have simulated, which have surface areas covering the range of 100 to >1500 $m^2\ g^{-1}$, to improve understanding of this relationship.

(629h) Computer-Assisted Screening of Ordered Crystalline Nanoporous Adsorbents for Separation of Alkane Isomers

Dubbeldam, D., University of Amsterdam
Krishna, R., University of Amsterdam
Calero, S., University Pablo de Olavide
Yazaydin, O., University of Surrey
Originally presented on: 11/7/2013 10:22:00 - 10:38:00

The separation of linear, mono-branched, and di-branched isomers of alkanes is of significant importance in the petrochemical industry, in particular for producing high-octane gasoline. Using computational techniques to screen a large variety of adsorbent materials, ZIF-77 emerges as the best adsorbent. It has the unique ability to fractionate isomer mixtures according to the degree of branching.

We have rationalized the adsorption behavior of hexane (and heptane) isomers and our screening study provides a thorough understanding. First, it is an illadvised endeavor to use large-pore MOFs for alkane separations. Above a certain threshold (depending on the alkane chain-length) the excess pore volume just goes to waste. Below the threshold for selectivity, both branching hierarchies are possible, but the one corresponding to the largest pore volume (the normal hierarchy) is probably most efficient, also because adsorption and diffusion selectivity go hand-in-hand. To properly evaluate a structure at least a 5-component mixture is needed, with both the 2,2-di-branched and the 2,3-di-branched included, as these have usually very different behaviors. The simulated breakthrough curves provide the best way of assessing the efficiency of a structure. The efficiency depends on a) the selectivity, and b) the accessible pore volume. CoBDP provides a very good compromise between the two, but the selectivity of ZIF-77 is so high that this sieve fractionates to yield individual pure components. The screening strategy we present(1) can be applied to other problems, such as, for example, separating alkane mixtures with C10–C18 hydrocarbons and xylene isomers.

(1) Dubbeldam, D., Krishna, R., Calero, S., Yazaydin, A.O., "Computer-Assisted Screening of Ordered Crystalline Nanoporous
Adsorbents for Separation of Alkane Isomers", *Angewandte Chemie International Edition,* 2012, 51, 11867-11871.

(629i) Branched Versus Linear Alkane Adsorption in Carbonaceous Slit Pores

Harrison, A.,
Krueger-Venus, J.,
Cracknell, R., Shell Global Solutions
Sarkisov, L., University of Edinburgh
Originally presented on: 11/7/2013 10:38:00 - 10:54:00

The presence of carbonaceous deposits on the internal surfaces of a spark ignition engine has been linked in the literature to impaired vehicle performance, as manifested by increased knocking, higher fuel consumption, higher emissions and other adverse effects. One of the proposed mechanisms is the adsorption and desorption of fuel components in the pores within the deposit. In this presentation we investigate this mechanism in more detail by considering single component adsorption of normal and branched alkanes in selected model slit pores representing the structure of the deposits. We further extend these studies to the binary mixture of normal heptane and isooctane, corresponding to a primary reference fuel blend. In particular, we show that in larger pores adsorption selectivity towards isooctane is about 1.2 on average throughout the pressure range. However, in the smaller 10Å pore selectivity towards isooctane can be in excess of three as a result of packing effects. These results are then placed in the context of engine performance issues.

(645a) Experiment and Theory for the Dew Point of Water in Compressed CO2

Meyer, C. W., NIST
Harvey, A. H., National Institute of Standards and Technology
Originally presented on: 11/7/2013 8:30:00 - 8:52:00

When compressing and transporting CO_2 for carbon capture and sequestration, it is important to know the water dew point as a function of temperature, pressure, and water content; this determines the degree to which it is necessary to dry the CO_2 in order to avoid condensation that can lead to corrosion in pipelines. This dew-point relationship is sensitive to the vapor-phase nonideality, on which there is limited theoretical information and only a few scattered experimental data at conditions relevant for pipeline transport. Dew-point data can also inform thermodynamic modeling of other parts of the process where H_2O and CO_2 are present.

To provide needed data, we have constructed an apparatus to measure the water mole fraction at saturation in a compressed gas. The apparatus consists of a saturator and the NIST gravimetric hygrometer (an instrument originally designed for work in humidity standards). The saturator humidifies the gas by bringing it into equilibrium with liquid water at a precisely controlled temperature and pressure. Then, the gravimetric hygrometer measures the water mole fraction of the humid gas with a relative uncertainty better than 0.1%. For CO_2, the apparatus can operate at pressures up to 5 MPa and at temperatures from approximately 1 °C to 85 °C.

We have measured dew-point data for several isotherms both above and below room temperature. Preliminary analysis indicates that these data are consistent with what would be expected based on recent theoretical predictions by Wheatley and Harvey, based on an *ab initio* potential-energy surface, of the interaction second virial coefficient between water and carbon dioxide.

(645b) Analytical and Pore-Network Calculations of Methane Saturation Resulting From Hydrate Dissociation in Porous Media

Tsimpanogiannis, I. N., National Center for Scientific Research
Lichtner, P., Los Alamos National Laboratory
Originally presented on: 11/7/2013 8:52:00 - 9:14:00

The capability of clathrate hydrates to store large amounts of gases has attracted significant attention from the scientific and industrial communities. Practical applications such as the storage and transportation of "energy-carrier" gases (methane, hydrogen), or "green-house" gases (carbon dioxide), mixture separation, and desalination, are among those of interest. Flow assurance, ocean-floor stability, and hydrates as a major source of well-drilling hazards are issues of major concern to the oil/gas industry [1].

The discovery of large amounts of natural gas-hydrate deposits in oceanic or permafrost sediments has rendered hydrates as a possible future energy source. Of interest to this study is the case of methane hydrate in porous media, as a source of methane for energy applications. For this purpose the hydrate needs to dissociate in order for the gas phase to be released, and subsequently to be produced through production wells. Methods that are considered in order to achieve the particular goal of hydrate dissociation include pressure reduction, thermal stimulation, hydrate-inhibitor injection or combinations of the above methods. The gas saturation that results from the dissociation of methane hydrate inside aqueous-saturated porous domains is a crucial parameter that determines if the resulting methane gas phase can be produced. To this purpose the gas phase needs either to form a percolating gas cluster that connects to the producing well, or reach the well, as a result of gas bubble mobilization due to viscous or buoyancy forces.

In this study we develop predictive tools for methane gas saturation that results from hydrate dissociation in porous media at different length scales, ranging from the single-pore scale, up to the pore-network scale [2]. Porous-medium overall, as well as, transversely-averaged, spatially specific, gas saturations are calculated. The analytical calculations are compared with the results obtained from detailed simulations using pore networks [3] and both results are found to be in very good agreement. Pore networks have been used successfully to capture the randomness of real porous media, and simulate processes involving flow, heat/mass transfer, and reactions in porous media.

References

[1] Sloan, E. D.; Koh, C. A. Clathrate Hydrates of Natural Gases, 3rd edition, Taylor & Francis. CRC Press, 2008.

[2] Tsimpanogiannis, I. N.; Lichtner, P. C. *J. Phys. Chem. C* **2013**, In press.

[3] Tsimpanogiannis, I. N.; Lichtner, P. C. *Phys. Rev. E* **2006**, *74*, 056303.

(645c) Exploring the Effect of Pore Morphology On Model Electrochemical Double-Layer Capacitor Systems

Phillips, K. A., North Carolina State University
Gubbins, K. E., North Carolina State University
Originally presented on: 11/7/2013 9:14:00 - 9:36:00

Electrochemical double-layer capacitors (EDLCs) have drawn much attention as energy storage devices in recent years due to their long cycle life as well as their ability to uptake and discharge energy very rapidly and to store more energy than typical capacitors. Molecular simulations are employed to study electrochemical properties of model EDLC systems. A variety of simple pore sizes, shapes, and configurations are studied, as well as model carbons, which reproduce structural features of disorder carbons used in EDLC experiments. Surface charge densities of 0 C/m^2, 5 C/m^2, 15 C/m^2, and 25 C/m^2 are applied to the carbon structure, which are representative of typical values for EDLCs. Solutions of 1M NaCl, NaF, and $NaBF_4$ are used as electrolytes in each system. We study the effect of surface charge on the electrochemical and structural properties of the electrolyte solution. Of particular interest are: capacitance of the model EDLC system, selectivity of ions to reside in various pore sizes, charge density of the electrolyte solution across the electrode system, and the solvation structure of ions in various confining geometries compared to the bulk solution between electrodes.

(645d) Thermodynamic Energy and Exergy Efficiency Limits of Photosynthesis

Silva, C., University of Pennsylvania
Seider, W. D., University of Pennsylvania
Lior, N.,
Originally presented on: 11/7/2013 9:36:00 - 9:58:00

With recent concerns about sustainable resources and environmental protection, growing attention has been focused upon biological sources of chemicals and fuels; however, to analyze such bioprocesses for commercial viability, the theoretical efficiencies must be known as an upper-bound on performance. Since almost all exergy contained in biomass originated from solar radiation, photosynthesis is the gateway to sustainable bioprocess development. Yet, despite the mechanism for photosynthesis being known for decades, there is still no well-established definition for its efficiency. The two types of studies that exist in the literature consider either the thermodynamic/physical effects of the process (evaporation, carbon dioxide sequestration, temperature changes) [1], [2], [3] and ignore the complex biochemical reaction mechanism, or the converse [4] [5]. The disjunction is due to the dichotomy that exists between biology and traditional thermodynamics. This study dissects the complex bio-processes involved in photosynthesis and analyzes their energy and exergy effects, bridging the gap and creating a clearer picture that is acceptable to both thermodynamicists and biologists. For the light reactions, reduction potentials were used to analyze the flow of excited, high-energy electrons through photosystem II and photosystem I, resulting in an exergy efficiency of 25%. For the dark reactions, free energies of formation were taken from the literature and used to calculate the Gibbs free energy and exergy changes; exergy changes due to concentration changes were assumed to be negligible for both the light and dark reactions. The resulting efficiency of the dark reactions was 89.5%. An exergy balance was formulated for water flows throughout the system to account for exergy destruction due to transpiration. Similarly, an exergy balance was constructed to account for losses due to ATPsynthase as the proton-motive force is used to synthesize ATP. Heuristic values for photorespiration, photo-degradation, and absorbance were taken from the literature and combined with the theoretical analysis above, yielding an overall photosynthetic efficiency of 1.9%. The overall efficiency shows good agreement with recent publications in the literature [1], [2]. Ultimately, the largest losses were due to poor absorbance and electron transfer through the photosystems.

References:

[1] Bisio G, Bisio A. Some Thermodynamic Remarks on Photosynthetic Energy Conversion. Energy Conversion Management 1998; 39(8):741-748.

[2] Petela R. An approach to the exergy analysis of photosynthesis. Solar Energy 2008; 82: 311-328.

[3] Reis HA, Miguel AF. Analysis of the exergy balance of green leaves. Int. J. Exergy 2006; 3(3): 231-237.

[4] Lems S, van der Kooi HJ, de Swaan Arons J. Exergy analyses of the biochemical processes of photosynthesis. Int. J. Exergy 2010; 7(3): 333-351.

[5] Chain RK, Arnon DI. Quantum Efficiency of Photosynthetic Energy Conversion, Proceeding of the National Academy of Science 1977; 74(8): 3377-3381.

(645e) Energetics and Mechanism of Desorption of Small Molecules From Cellulose Crystal Surface: Atomic Force Microscopy (AFM) and Molecular Simulation Study

Muthukumar, L., Texas Tech University
Liao, Y. C.,
Weeks, B. L., Texas Tech University
Khare, R., Texas Tech University
Originally presented on: 11/7/2013 9:58:00 - 10:20:00

The cello-oligosaccharide intermediates formed during enzymatic hydrolysis of cellulose are adsorbed on the cellulose crystal surface thus hindering further action of the enzyme. In this work, we have used a combination of AFM experiments and molecular modeling to decipher both the energetics and mechanism of the process of desorption of cello-oligosaccharide molecules from the cellulose crystal surface. For this purpose, molecules of model compounds containing different functional groups are attached to the AFM tip and the adhesion force between these molecules and the cellulose crystal surface is determined using AFM experiments. The steered molecular dynamics (SMD) technique is used to determine the adhesion force as well as the mechanism of desorption of these same molecules from the cellulose crystal surface. Simulation results obtained for different crystal surfaces as well as for varying number of adsorbed molecules are directly compared with AFM experiment results.

(645f) Modelling Super-Equilibrium in Biomass Gasification With the Constrained Gibbs Energy Method

Kangas, P., VTT Technical Research Centre of Finland
Koukkari, P., VTT Technical Research Centre of Finland
Hannula, I.,
Originally presented on: 11/7/2013 10:20:00 - 10:42:00

Background

Biomass gasification can be seen as a viable option for generating electricity from biomass with high efficiency or for the production of liquid biofuels and chemicals. Gasification is a thermo-chemical conversion process that takes place at high temperature and reductive environment. For such conditions thermodynamic equilibrium is often assumed, but not reached in practice: during biomass gasification light hydrocarbons (e.g.: CH_4, C_2H_2, C_2H_4, C2H6, C_3H_8, C_6H_6), ammonia, tars and char are also formed. These phenomena cannot usually be modelled with equilibrium assumptions.

In order to increase the accuracy of gasification models, both empirical and mechanistic models are often applied alone or combined with thermodynamic equilibrium calculations [1]. This leads to a 'dual' approach where different parts of the process and reactions are modelled independently and results are merged in post-processing.

This study presents a unified solution for simultaneous calculation of the super-equilibrium reactions and related reaction enthalpies in the gasification process. The solution is based on the Constrained Free Energy (CFE) method where equilibrium computation is extended with additional immaterial constraints for solving the local or partial constrained equilibrium instead of the global thermodynamic equilibrium.

Methodology

First applications of the constrained thermodynamic equilibrium method in Gibbs energy minimization were the conservation of immaterial properties [2] and inclusion of reaction kinetics into the Gibbs'ian multicomponent models [3,4]. In the course of time the method has found new applications, ranging from modelling sorption in pulp suspensions to determining surface tension of liquid mixtures [5]. Quite recently CFE has been used to estimate the super-equilibria of alkali metals and sulphur in black liquor combustion [6]. Today constrained free energy method is seen as a versatile tool for various fields of application.

In this study the CFE methodology is applied for describing the super-equilibrium occurring in biomass gasification. Light hydrocarbons, ammonia, tars and char tend to decompose if thermodynamic equilibrium calculation is performed for high temperatures, and thus constraints are needed for the modelling of their presence in the super-equilibrium conditions.

This a computational study with all experimental data obtained from literature. The empirical model of super-equilibrium applied is from [1]. Thermodynamic data is from [7]. ChemSheet [8] is used as the modelling tool as it allows extending the thermodynamic system and thus enables the calculations of constrained free energy models.

Results

In the present work different approaches for the modelling of global or local equilibrium of biomass gasification have been assessed. First (i) thermodynamic equilibrium without any constraints is computed and used as reference. Later additional constraints are introduced to system, such as amount of (ii) char, (iii) tars, (iv) ammonia. The light hydrocarbons are defined by (v) the amount of carbon and by (vi) amount of hydrogen in hydrocarbons. In addition constraints such as (vii) amount of double bonds in hydrocarbons or (viii) methane amount are evaluated. Finally a fully constrained system, where all species are defined according to [1] is reported.

The super-equilibrium of major syngas components (CO, CO_2, H_2, H_2O and CH_4) can be predicted with the proposed constrained free energy method. Accuracy of the model increases when an additional constraint for hydrogen in hydrocarbons is defined. At the same time the partial equilibrium moves further from the global equilibrium.

While the adapted constraints were successful in evaluating the super-equilibrium of major species, the respective predictions for minor hydrocarbon components (C_2H_2, C_2H_4, C_2H_6, C_3H_8, C_6H_6) were not as satisfying. The selection of modelled light hydrocarbon components seems to be vital, if more accuracy is required. In addition by defining all species with separate constraints, it is possible to increase the accuracy of results, but with the cost of decreasing the degrees of freedom of the thermodynamic system.

Summary

Constrained free energy method was successfully applied for the calculation of the super-equilibrium of light hydrocarbons, ammonia, tars and char in biomass gasification. Gasification can be solved as restricted partial equilibrium with a single calculation step. Thus, there is no need for post-processing corrections for e.g. volume or enthalpy flows. For a reasonable accuracy of the simulation model, it is not necessary to define constraints for all reactions, but a set of more general constraints (e.g. carbon in hydrocarbons) remain sufficient.

Distinct benefit of presented approach is that chemical reactions, gasification enthalpy and states of the system (such as volume) are estimated concurrently. This method is also resilient: the number of details describing constraints can vary from simpler experimental results and models to complicated mechanistic models. A promising application area for the presented methodology could be as part of large scale process simulation where detailed thermodynamic models can be implemented and effects in gasification chemistry could conduct changes in the entire process.

References

[1] Hannula, I. and Kurkela, E., "A parametric modelling study for pressurised steam/O2-blown fluidised-bed gasification of wood with catalytic reforming", Biomass and Bioenergy, 38:58–67 (2012).

[2] Alberty, R.A., "Thermodynamics of the formation of benzene series polycyclic aromatic

hydrocarbons in a benzene flame", The Journal of Physical Chemistry, 93(8):3299–3304 (1989).

[3] Keck, J.C., "Rate-controlled constrained-equilibrium theory of chemical reactions in complex systems", Progress in Energy and Combustion Science, 16(2):125–154 (1990).

[4] Koukkari, P., "A physico-chemical method to calculate time-dependent reaction mixtures", Computers & Chemical Engineering, 17(12):1157–1165 (1993).

[5] Koukkari, P. and Pajarre, R., "Calculation of constrained equilibria by Gibbs energy minimization", Calphad, 30(1):18–26 (2006).

[6] Kangas, P., Koukkari, P., Lindberg, D. and Hupa, M., "Modelling black liquor combustion with the constrained Gibbs energy method", Proceedings, 8th International Black Liquor Colloquium, Belo Horizonte, Brazil (2013).

[7] Roine, A., Lamberg, P., Mansikka-aho, J., Björklund, P., Kentala, J., Talonen, T., et al., HSC Chemistry 6.12., Outotec Research Oy, (2007).

[8] Koukkari, P., Penttilä, K., Hack, K. and Petersen, S., "CHEMSHEET – An Efficient Worksheet Tool for Thermodynamic Process Simulation", In: Microstructures, Mechanical Properties and Processes, Y. BrÃ©chet (Ed.), Wiley-VCH Verlag GmbH & Co. KGaA, Weinheim (2005).

(645g) Microscopic Insights Into the Electrochemical Behavior of Non-Aqueous Electrolytes in Electric Double-Layer Capacitors

Wu, J., UC Riverside
Jiang, D. E., Oak Ridge National Laboratory
Originally presented on: 11/7/2013 10:42:00 - 11:04:00

Electric double-layer capacitors (EDLC) are electrical devices that store energy by adsorption of ionic species at the inner surface of porous electrodes. Compared with aqueous electrolytes, ionic liquid and organic electrolytes have the advantage of larger potential windows, making them attractive for the next generation of EDLC with superior energy and power densities. The performance of both ionic liquid and organic electrolyte EDLC hinges on the judicious selection of the electrode pore size and the electrolyte composition that requires a comprehension of the charging behavior from a microscopic view. In this perspective, we discuss predictions from the classical density functional theory (CDFT) on the dependence of the capacitance on the pore size for ionic-liquid and organic-electrolyte EDLC. CDFT is applicable to electrodes with the pore size ranging from that below the ionic dimensionality to mesoscopic scales, thus unique for investigating the electrochemical behavior of the confined electrolytes for EDLC applications.

(667a) Effect of Langmuir Constants On Hydrate Equilibrium Calculations

Tsimpanogiannis, I. N., National Center for Scientific Research
Economou, I. G., National Center for Scientific Research â€œDemokritosâ€
Stubos, A. K., National Center for Scientific Research "Demokritos"
Originally presented on: 11/7/2013 12:30:00 - 12:47:00

Under appropriate conditions of pressure and temperature water molecules can self-assemble to form cavities that are stabilized by the presence of guest molecules that can fit within the cavities. These non-stoichiometric, solid (ice-like), inclusion materials are known as clathrate hydrates and have significant applications in science and engineering. Hydrate equilibrium at the continuum (i.e., macroscopic) level is described by a theory based on statistical mechanics that has been developed, initially by van der Waals and Platteuw [1], and subsequently modified by others (e.g., see the extended review by Sloan and Koh [2]). In this type of theory the cavity occupancies (i.e., the fraction of cavities occupied by the guest molecules) are described by Langmuir-type functions of the gas fugacity. The Langmuir constants that appear in these functions are temperature, guest, as well as, cavity-type-dependent, and affect the accuracy of the calculations. Therefore, it is essential to develop dependable and accurate methods for the calculation of the Langmuir constants. Traditionally, two approaches are followed in order to calculate the Langmuir constants:

(i) An intermolecular potential is chosen to describe the water-guest interactions (e.g., the Lennard-Jones, the square-well, or the Kihara potential), and the potential parameters are obtained by a fitting procedure using available experimental hydrate equilibrium data. Following this approach results in very good agreement with hydrate equilibrium predictions (within the range of development of the method), however, the values for the potential parameters are significantly different than those calculated when using experimental data of second virial coefficients or transport properties to obtain the potential parameters. Therefore, extrapolation to hydrate equilibrium predictions outside the range of development becomes highly problematic.

(ii) The interaction energies between the guest molecules and the water molecules that form the hydrate cavities are calculated using ab initio methods [3]. The ab initio calculation results, subsequently, are fitted to an empirical formula (e.g., the Lennard-Jones plus a Coulombic charge-charge formula) in order to obtain the potential parameters [4]. The accuracy of calculations of ab initio potential energy surfaces depends on the level (e.g., HF, MP2, MP3, MP4, etc.) of the simulations, and the basis sets used. Increasing the complexity of the calculation, results in increasing significantly the computational cost.

Alternatively, the following approach could be used. Hydrate formation can be simulated as a special case of a process of gas adsorption in a solid matrix. Such processes have been simulated with Monte Carlo techniques with great success. The hydrate cavity occupancies, thus, could be calculated from the Monte

Carlo simulations at various temperatures and pressures, and the results can be fitted to Langmuir-type functions [5]. Such temperature and pressure-dependent functions, once they are obtained, can be introduced to models that use the van der Waals and Platteuw theory for hydrate equilibrium predictions. However, the hydrate equilibrium predictions of such macroscopic models seem to be very sensitive to the values of the Langmuir constants used. As a result of such sensitivity, other constants of the macroscopic thermodynamic models need to be re-adjusted in order to increase the accuracy of the calculations. In the current study we perform a detailed analysis of the sensitivity of the hydrate equilibrium calculations on the values used for the Langmuir constants. We attempt to delineate the boundaries of the values of the Langmuir constants obtained from Monte Carlo simulations that can be used by van der Waals–Platteuw–type models without having to re-adjust any other of the model parameters.

References

[1] van der Waals, J. H.; Platteeuw, J. C. *Adv. Chem. Phys.* **1959**, *2*, 1-57.

[2] Sloan, E. D.; Koh, C. A. Clathrate Hydrates of Natural Gases, 3rd edition, Taylor & Francis. CRC Press, 2008.

[3] Klauda, J. B.; Sandler, S. I. *J. Phys. Chem. B* **2002**, *106*, 5722-5732.

[4] Sun, R.; Duan, Z. *Geochim. Cosmochim. Acta* **2005**, *69*, 4411-4424.

[5] Papadimitriou, N. I.; Tsimpanogiannis, I. N.; Papaioannou, A. Th.; Stubos, A. K. *J. Phys. Chem. B* **2008**, *112*, 10294-10302.

(667b) Self-Assembly of Aqueous Surfactants On Patterned Surfaces: Insights From Coarse-Grained Simulations

Suttipong, M., The University of Oklahoma
Grady, B. P., University of Oklahoma
Striolo, A., University of Oklahoma
Originally presented on: 11/7/2013 12:47:00 - 13:04:00

The adsorption and self-assembly of surfactants are ubiquitous processes in several technological applications. From a fundamental point of view, understanding such processes is also important because it can have profound implications in the manufacture of nano-structured materials using bottom-up strategies. Although a lot is known about the adsorption of surfactants on homogeneous flat surfaces, limited information is available, in quantifiable terms, regarding the adsorption of surfactants on heterogeneous surfaces. To complement and enrich our recent experimental observations [Langmuir 27 (2011) 6091], we report here our first results obtained using dissipative particle dynamics simulations for the adsorption of model surfactants on patterned flat surfaces. The results show that the surface pattern can be used to direct the morphology of the self-assembled aggregates (flat monolayers, hemi-cylinders, hemi-spheres, hollow hemi-cylinders, and irregular structures have all been observed using the same surfactant but different surface patterns). We will discuss how our predictions could be tested experimentally, and how, if reliable, they could be used for the design of new materials.

(667c) Equations of State for Hard Polyhedra

Irrgang, M. E., University of Michigan Ann Arbor
Engel, M., University of Michigan
Glotzer, S. C., University of Michigan
Originally presented on: 11/7/2013 13:04:00 - 13:21:00

The role of shape in entropy-driven self-assembly has recently been highlighted in computer simulations of hard anisotropic particles. A rich diversity of crystal and other solid-like phases has been demonstrated in particular for hard polyhedra [1-5]. Hard polyhedra and other simple shapes can be good models for nano crystalline colloids when interactions are dominated by geometry. Though advancements in the synthesis of such nanoparticles[7] are expanding our ability to create novel fluids and structured hierarchical materials, our understanding of the thermodynamics of anisotropic particles is still limited. Here we present equations of state for systems of hard polyhedra spanning the low-density fluid to high-density solid states, obtained numerically from equilibrium Monte Carlo simulations. We compare the results to various forms of equations of state from free volume and scaled particle theory. We discuss trends in the behavior for different shapes, and show some general features common to all systems.

[1] Haji-Akbari, A. et al., 2009. Disordered, quasicrystalline and crystalline phases of densely packed tetrahedra. Nature, 462(7274), pp.773–7.

[2] Agarwal, U. & Escobedo, F. A., 2011. Mesophase behaviour of polyhedral particles. Nature Materials, 10(3), pp.230–235.

[3] Ni, R. et al., 2012. Phase diagram of colloidal hard superballs: from cubes via spheres to octahedra. Soft Matter, 8(34).

[4] Damasceno, P.F., Engel, M. & Glotzer, S.C., 2012. Predictive Self-Assembly of Polyhedra into Complex Structures. Science, 337(6093), pp.453–457.

[5] Damasceno, P.F., Engel, M. & Glotzer, S.C., 2012. Crystalline assemblies and densest packings of a family of truncated tetrahedra and the role of directional entropic forces. ACS Nano, 6(1), pp.609–14.

[6] Henzie, J. et al., 2012. Self-assembly of uniform polyhedral silver nanocrystals into densest packings and exotic superlattices. Nature Materials, 11(2), pp.131–7.

(667d) Formation of Interconnected Morphologies of Symmetrical Block Copolymer/Nanorod Comoposites Under Cylindrical Confinement: A Coarse-Grained Molecular Dynamics Study

Park, J. H., Cornell University
Joo, Y. L., Cornell University
Originally presented on: 11/7/2013 13:21:00 - 13:38:00

Our recent experiment and simulation on confined self-assembly of block copolymer (BCP)/nanoparticle (NP) composites in electrospun nanofiber have proven to be a useful guidance for fabrication of the composite materials with ordered nanostructures. Specifically, we have examined how nanoparticle presence can influence the formation of very different asymmetrical BCP morphologies under cylindrical confinement when compared with that of pure asymmetrical BCP, both qualitatively and quantitatively. Using the same coarse-grained molecular dynamics tool, we have further extended our study to self-assembly of nanorods with BCP under cylindrical confinement. Unlike nanoparticles, nanorods can readily be interconnected with each other, and also induce asymmetrical BCP morphologies at much lower loading that nanoparticles do. Such interconnected nanorods, when incorporated within nanofibers, have great potential to provide highly conductive pathways for energy applications, such as battery electrodes and separators. Symmetric BCP under cylindrical confinement with nanorod aspect ratio of 1, 5, and 10 are examined with three different types of nanorod-BCP attraction: a) neutral nanorod, b) *A* (wall-attractive phase)-attractive nanorods, and c) *B* (wall-repulsive phase)-attractive nanorods. As one may have expected, much less loading of longer nanorods are required to observe formation of radially and axially interconnected nanorods within the cylinder. The interconnected nanorods often, but not always, induce asymmetrical BCP morphology as well. From this trend, we were able to extract a dimensionless parameter, *D*/*N*, or the polymer domain diameter over the nanorod aspect ratio, to be correlated to the onset of nanorod interconnectivity and formation of asymmetrical BCP morphology. By developing a phase diagram with respect to *D*/*N*, one can predict the formation of desired BCP morphology, nanorod placement, and it connectivity within cylindrical confinement.

(667e) Parameterization of Coarse-Grained Models for Simulation of Self-Assembly: the Role of Molecule Rigidity

Lee, M. T., Rutgers, The State University of New Jersey
Vishnyakov, A., Rutgers, The State University of New Jersey
Neimark, A. V., Rutgers, The State University of New Jersey
Originally presented on: 11/7/2013 13:38:00 - 13:55:00

Micellization of surfactant solutions is a ubiquitous phenomenon in natural systems and technological processes, and its theoretical description is a cornerstone problem in physical chemistry of colloidal systems. However, successful attempts of quantitative modeling confirmed by experimental data remain limited. We devised a novel approach to parameterization of coarse-grained models: binary interaction parameters are fitted to the infinite dilution activity coefficients of binary solutions formed by reference compounds that represent coarse-grained fragments of surfactant molecules. Using this protocol, we derived dissipative particle dynamics (DPD) models of several surfactants of different chemical structures and obtained a quantitative agreement with experimental critical micelle concentration (CMC) and aggregation number (AN).

Another factor determining micellization in surfactant solutions is the rigidity of the molecules. For the first time, we present a systematic study of chain rigidity effect on CMC and AN using DPD simulations. Molecule rigidity was controlled by second-neighbor (?1-3?) harmonic bonds, or the harmonic angle potential between the nearest neighbor bonds. Compared to flexible molecules with the nearest neighbor bonds being the only type of bonded interactions, rigid molecules exhibited a lower critical micelle concentration, larger and better-defined micelles. By varying the strength of head-tail repulsion and chain stiffness, we constructed two-dimensional diagrams presenting how the critical micelle concentration and aggregation number depend on these parameters. We found that solutions of flexible and rigid molecules that exhibited approximately the same critical micelle concentration could differ substantially in the micelle size and shape depending on the chain stiffness. With the increase of surfactant concentration, primary micelles of more rigid molecules were found less keen to agglomeration and formation of non-spherical aggregates characteristic to flexible molecules.

Overall, our parameterization methodology can be recommended for studies of self-assembly and dynamics in a wide range of soft matter systems. However, rigidity has to be accurately accounted for in coarse-grained modeling of self-assembly.

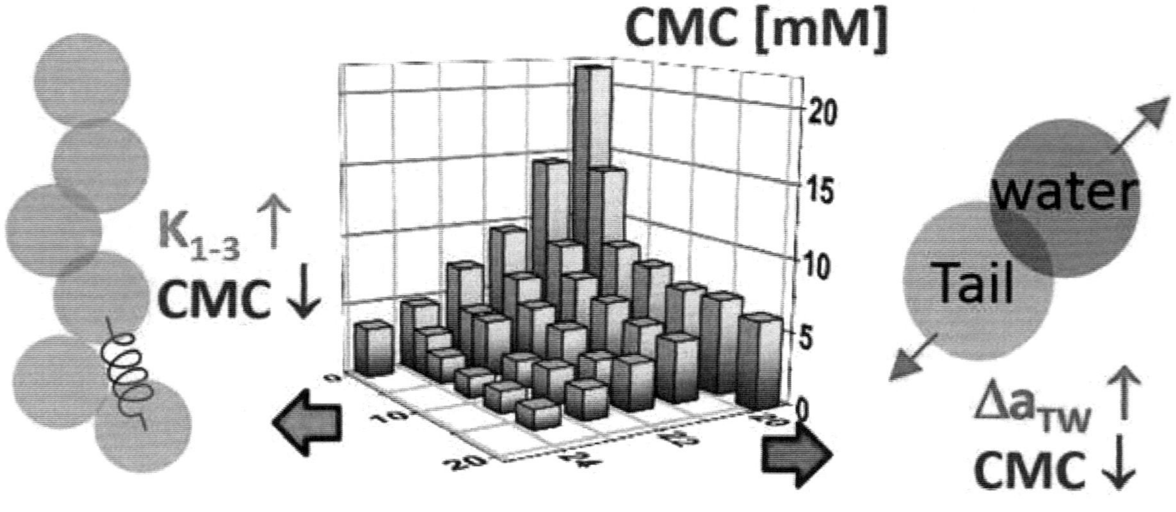

(667f) New Theory for Associating Fluids: Effects of Steric Hindrance, Ring Formation and Double Bonding

Marshall, B. D. II, Rice University
Chapman, W. G., Rice University
Originally presented on: 11/7/2013 13:55:00 - 14:12:00

Equations of state based on Wertheim's first order perturbation theory (TPT1) have proven to be a powerful tool in the modeling of associating fluids. In TPT1 the effect of bond angle is neglected such that the equation of state for associating molecules with a bond angle of 60 degrees is identical to that for associating fluids with bond angles of 180 degrees; that is, the effect of steric hindrance between association sites is neglected. Also neglected in TPT1 is the ability for molecules to associate into rings and double bond. For large bond angles these assumptions are valid; however, we show through molecular simulation that as bond angles are decreased steric hindrance between sites, association into rings and the possibility of double bonding must all be accounted for.

To model these small bond angle systems we develop a new molecular theory which captures each one of these small bond angle effects (steric hindrance, rings, double bonds). We explicitly include the effect of bond angle in each contribution to the free energy to develop an equation of state which is accurate over the full range of bond angles. We show by comparison to molecular simulation that the theory accurately predicts the pressure, internal energy and distributions of molecules in the various types of associated clusters (chains, double bonded, 3-mer rings, 4-mer rings, etc…). Further, we use the theory to demonstrate the effect of bond angle on the phase diagram of patchy colloid fluids.

(667g) Prediction of Phase Behavior of Strongly Associating Fluids By Explicit Consideration of Transitions in Local Fluid Structures

Chen, W. L., National Taiwan University
Hsu, C. C., National Taiwan University
Lin, S. T., National Taiwan University
Originally presented on: 11/7/2013 14:12:00 - 14:29:00

Accurate modeling of phase behavior of acetic acid has been of great interest both in academia and industry. The challenge is to accurately describe the various local structures (hydrogen bonding configurations) and the transition in the distribution of local structures associated with phase transitions. In this work we use the PR+COSMOSAC EOS to predict the phase behaviors of fluids containing acetic acid. The PR+COSMOSAC EOS does not use any species dependent parameter and does not require input of experimental data. However to accurately model strongly associating fluids, it is necessary to identify the important local fluid structures such as the monomer, cyclic dimer, chain fragment of acetic acid and the cross dimer of acetic acid and alcohol. The transformations among these structures are considered by chemical reactions. With the explicit inclusions of these associating structures the PR+COSMOSAC EOS is as accurate as the mod-UNIFAC+ Hayden-O'Connell EOS in describing the vapor liquid equilibrium of 19 binary mixtures (447 data points, temperature range from 293.15 to 426.39) is 6.49% (AARD-P%) in pressure and 2.22% (AAD-y%) in vapor composition, which is similar to those from the mod-UNIFAC+HOC 5.16% and 1.87%. The root-mean square error in predicting liquid composition in liquid-liquid equilibrium of 5 binary systems (51 data points, temperature range from 289.15K to 337K) is found to be 0.091, which is more accurate than that from mod-UNIFAC, 0.173. Our results show that the phase behavior of strongly associating systems can be predicted accurately by explicit consideration of the significant local fluid structures.

(667h) Molecular Insights Into Diphenylalanine Nanotube Assembly: All-Atom Simulations of Oligomerization

Jeon, J. H., University of California Santa Barbara
Mills, C. E., University of California Santa Barbara
Shell, M. S., University of California, Santa Barbara
Originally presented on: 11/7/2013 14:29:00 - 14:46:00

The simple diphenylalanine (FF) peptide self-assembles in aqueous solutions into nanotubes (FFNTs) with remarkable strength, thermal stability and other promising physical properties. FFNTs have found use in many applications, including as sacrificial templates and scaffolds for structuring inorganic materials, and as new hydrogels and biosensors. FFNTs can also form on substrates using vapor-deposition in vacuum, to create dense FF-nanoforests that have already found use in super-hydrophobic surfaces and high-performance supercapacitors. However, little is known about assembly mechanisms of FFNTs or the forces underlying their stability.

Here, we use detailed molecular simulations to study the early self-assembly stages of FF in water and in vacuum. First, we perform a variety of molecular dynamics (MD) simulations on small-oligomer formation in water and assess the balance of hydrogen bonds, electrostatic interactions, and side chain aromatic or hydrophobic forces, as well as the emergent structural motifs. We find that while electrostatic interactions steer FF peptides into more ordered dimers and trimers, the hydrophobic side chain interactions play a strong role in determining the structures of larger oligomers. However, capped (uncharged) versions of the peptide display dramatically different assembly behavior, and emphasize the importance of amphiphilicity to FF assembly. By comparing these results to simulations of the experimental FFNT X-ray crystal structure, we propose that the early structural templates to be formed are more amphiphili-like, rather than the salt-bridge-stabilized hexamer ring motif that has been proposed in the literature.

Second, we study the assembly of FF peptides in vacuum. During the vaporization process, the linear backbone of FF peptides is cyclized, and so its chemical structure is distinct from that in solution. We find that cyclo-FF molecules indeed have strong preferences to form vertically aligned ladder-like structures stitched together by hydrogen bonds. Such structures are reminiscent of the solution-phase oligomers, but appear much more stable. Importantly, the obtained ladder structures have some structural similarities with the crystal structure of cyclo-FF nanowires, although future work mush address how these ladders pack in a way to form nanotubes. In total, our results suggest novel pathways for assembly of FF in different environments and offer plausible explanations for the interactions that drive their unique properties.

(667i) Effect of Solvent Additives On the Self-Aggregating Properties of Phenyl-C61-Butyric Acid-Methyl-Ester (PCBM)

Tummala, N. R., The University of Oklahoma
Risko, C., Georgia Institute of Technology
Bredas, J. L., Georgia Institute of Technology
Originally presented on: 11/7/2013 14:46:00 - 15:03:00

In organic photovoltaics, the morphology at the interface between the electron-rich (donor) and electron-deficient (acceptor) materials in a bulk hetero-junction (BHJ) film directly impacts the critical processes of exciton dissociation and charge separation; on the other hand, aggregates of the individual components influence the charge-carrier mobilities in the bulk-like areas of the active layer. Solvent additives with high boiling points have been found to improve the efficiency of organic solar cells[1, 2] by aiding in the formation of improved morphological or topological[3-5] features within the active layer of the BHJ. Here, we aim to understand how solvent additives induce self-aggregation in solution, prior to spin-casting, through molecular dynamics simulations with force fields previously validated within our group. We evaluate properties like aggregation numbers, energies of aggregates of different sizes, cluster-size distributions in various solvents, and cluster shapes. By understanding the aggregation of PCBM clusters in chlorobenzene with and without solvent additives (*e.g.*, 1-chloronaphthalene and 1-8-diiodooctane), we determine the role of solvent additives in the self-aggregation processes and PCBM-PCBM orientations. We also evaluate the energetic landscape of a PCBM dimer in multiple solvent / solvent additive environments to investigate how local molecular packing affects the ordering of small-to-large clusters. These studies shed light on the morphological nature of the PCBM domains, an important feature that affects the power conversion efficiency of organic solar cells.

References

[1] J. K. Lee, W. L. Ma, C. J. Brabec, J. Yuen, J. S. Moon, J. Y. Kim, K. Lee, G. C. Bazan, A. J. Heeger, J. Am. Chem. Soc. 2008, 130, 3619.

[2] J. S. Moon, C. J. Takacs, S. Cho, R. C. Coffin, H. Kim, G. C. Bazan, A. J. Heeger, Nano Lett. 2010, 10, 4005.

[3] F. Etzold, I. A. Howard, N. Forler, D. M. Cho, M. Meister, H. Mangold, J. Shu, M. R. Hansen, K. Müllen, F. Laquai, J. Am. Chem. Soc. 2012, 134, 10569.

[4] L. Ye, S. Zhang, W. Ma, B. Fan, X. Guo, Y. Huang, H. Ade, J. Hou, Adv. Mater. 2012, 24, 6335.

[5] X. Liu, S. Huettner, Z. Rong, M. Sommer, R. H. Friend, Adv. Mater. 2012, 24, 669.

(675a) Electron-Hole Correlation in Quantum Wires

Chakraborty, A., Syracuse University

Originally presented on: 11/7/2013 12:30:00 - 12:50:00

Indium gallium nitride is a very promising material for applications in visible light emitting devices. This presentation will focus on computational treatment of the underlying electron-hole recombination process in InGaN/GaN nanowire. The explicitly correlated configuration interaction (XCCI) method will be used for solving the electron-hole Schrodinger equation for nanowires and electron-hole interaction energy and electron-hole recombination probability will be computed. The XCCI method is a variational method and has been used successfully for computation of exciton binding energy, electron-hole recombination probability and quantum confined Stark effect in quantum dots. The two areas this talk will focus on are (1). Shape-bases control and (2). Heterojunction-based control of electron-hole recombination in nanowires. For shape-based control studies, the XCCI method was applied to a series on nanowire of different lengths and aspect ratios. A detailed comparison of recombination probabilities from these calculations will be presented. For heterojunction-based studies, effect on the electron-hole recombination probability due to the size of the core and its orientation with respect to the longitudinal axis of the nanowire (centered versus off-centered) will be presented. In addition to computation of electron-hole interaction energy and electron-hole recombination probability, average electron-hole interparticle distance and spatial distribution of the charge carriers will also be presented for the systems described above. The implication of the computational results to enhance efficient of the nanowires will be discussed.

(675b) Chemistry of Iron-Sulfur Containing Enzymes

Sharma, S., Massachusetts Institute of Technology
Chan, G.,
Originally presented on: 11/7/2013 12:50:00 - 13:10:00

Metals in enzymes perform remarkable chemistry under ambient pressures and temperatures. Among the most common metal cofactors are the iron-sulfur (FeS) clusters, containing between one and eight Fe atoms bridged by S ligands. FeS clusters perform multiple roles in nature, including redox chemistry, electron transfer, and even oxygen sensing, in processes ranging from nitrogen fixation to photosynthesis and respiration.

FeS cluster have multiple low-lying states with differing electronic and magnetic character, which are the key to their rich chemistry. However, determining the details of these states is a highly non-trivial problem. Traditional approaches like Density Functional Theory (DFT), although cheap and usually quite accurate, fail spectacularly when applied to transition metal containing systems. With recent advances in electronic structure theory, in particular with the advent of Density Matrix Renormalizaton Group (DMRG), we now have a method that is able to correctly describe the challenging chemistry of these enzymes.

In this work we will summarize the results of our DMRG calculations on Fe_2S_2 and Fe_4S_4 containing model complexes. Our detailed calculations reveal a dense spectrum of electronic states that intriguingly suggests that these complexes behave more like heterogeneous catalysts than simple molecular catalysts. This has led us to revise the canonical models that are the basis of our understanding of the chemistry of these complexes.

(675c) Optimization of a Reactive Force Field for Modeling the Polycondensation of Siloxanes

Deetz, J. D., UC Davis
Faller, R., UC Davis
Originally presented on: 11/7/2013 13:10:00 - 13:30:00

We investigate the structures of silicates obtained by successive hydrolysis and condensation reactions of siloxanes using Molecular Dynamics simulations. To characterize the system accurately, the ReaxFF force field previously used to describe the thermal decomposition of polysiloxanes was extended to model the bulk condensation of siloxanes. The parameter set was refined by using multiple algorithms. Both ab-initio calculations and bulk structure information are used to develop the model. Density functional theory calculations compute the energies of several locations along the reaction coordinate for hydrolysis and condensation reactions conditions. Geometric data such as bond lengths and angles, as well as the energies of their distortion are used. In order to accurately model the electrostatics of the system, point charges are derived in DFT using an electrostatic potential method (CHELPG). We also use experimental data for bulk phases, such as the enthalpy of vaporization, liquid density, and crystal lattice parameters, to constrain the long range forces during the optimization. A reasonable fit is found to available data, and a useful tool for modeling the gelation of silanes has been developed.

(675d) Virial Coefficients Of Hydrogen and Nitrogen Including Quantum Effects Using Path Integral Monte Carlo

Subramanian, R., State University of New York, University at Buffalo
Schultz, A. J., University at Buffalo, The State University of New York
Kofke, D. A., University of Buffalo
Originally presented on: 11/7/2013 13:30:00 - 13:50:00

Accurate virial coefficients are important for predicting properties of light gases such as hydrogen, helium etc., especially at low temperatures where the quantum effects aren't negligible. In addition to predicting properties of gases, virial coefficients (particularly the second virial coefficient) can be utilized as a tool to check the accuracy of various potential models against experimental data. Hence, the accuracy of such coefficients are of vital importance to many scientists and fellow engineers. The Mayer-sampling Monte Carlo method [1] provides an efficient route to calculating accurate virial coefficients. To incorporate quantum effects of light gases in which the positions of the atoms/molecules fluctuate by a significant amount at low temperatures, one needs to use path integral techniques. Path Integral Monte Carlo (PIMC) methods involve discretizing the quantum fluctuations of the atoms/molecules into "beads" that form closed paths or rings. Neighboring beads act as if they are harmonically connected to each other and the strength of the harmonic spring is directly proportional to temperature. The interaction energy is then computed as the average of the potential evaluated between corresponding beads of different rings. At very low temperatures, a large number of beads are required to accurately represent the path of the atoms/molecules.

PIMC methods have been applied by us and others to study quantum virial coefficients of He [2 - 5] and H_2 [6]. We now extend this approach to compute higher order virial coefficients using flexible potentials (if available) of small diatomic molecules such as H_2 and N_2. For diatomic molecules, the complexity in terms of the orientational sampling under the approximation of quantum rigid rotors involved in the PIMC methods can be greatly reduced by a new algorithm we introduce. We studied the probability distribution of placing each successive bead on the surface of a sphere as a function of the angle φ between the previous bead and the current bead position vectors. We noticed that for reasonably low φ, ($\varphi \sim< 15-20$ degrees) we could analytically express the probability distribution functions for all the beads as one universal probability distribution function with a parameter k, that was different for each bead. The analytical nature of this universal probability distribution function is the key feature that makes the algorithm fast, as it can be computed on the fly. Also, since we are regrowing the entire ring from scratch for each Monte-Carlo move, the possibility of different starting points of the these rings is accounted for. The assumption of low φ is reasonable even at low temperatures where the harmonic springs become weak. We present results for the H_2 and N_2 systems demonstrating the importance of the quantum effects, and providing a comparison with experimentally-obtained virial coefficients, where available.

[1] – Jayant K. Singh, David A. Kofke., Phys. Rev. Lett. 92 (22), 220601 (2004),DOI:10.1103/PhysRevLett.92.220601

[2] – Giovanni Garberoglio, Allan H. Harvey, J. Res. Natl. Inst. Stand. Technol. 114, 249 (2009), DOI:10.6028/jres.114.018

[3] – Giovanni Garberoglio, Allan H. Harvey, J. Chem. Phys. 134, 134106 (2011), DOI:10.1063/1.3573564

[4] – Giovanni Garberoglio, Micheal R. Moldover, Allan H. Harvey, J. Res. Natl. Inst. Stand. Technol. 116, 729 (2011), DOI:10.6028/jres.116.016

[5] – Katherine R.S. Shaul, Andrew J. Schultz, David A. Kofke, J. Chem. Phys. 137, 184101 (2012), DOI:10.1063/1.4764857

[6] – Giovanni Garberoglio, Chem. Phys. Lett. 557, 26 (2013), DOI:10.1016/j.cplett.2012.11.090

(675e) First Principles Thermodynamics of Amorphous Silica

Ewing, C. S., University of Pittsburgh
Veser, G., University of Pittsburgh
McCarthy, J. J., University of Pittsburgh
Johnson, J. K., University of Pittsburgh
Originally presented on: 11/7/2013 13:50:00 - 14:10:00

Silica nanoparticles have a wide range of potential applications including drug delivery, imaging, electronics, and catalysis. These applications typically utilize large systems of nanoparticles, thus their collective properties are important for optimizing system properties for specific applications. A detailed description of the amorphous silica surface at the atomic level is crucial to developing a complete understanding of the interactions between silica nanoparticles. While properties of silica surfaces have been heavily investigated over past years, an accurate atomistic description is lacking due to the complexity of the amorphous surface.

We have used a novel approach to obtain a detailed atomistic description of the amorphous silica surface, with a central focus on thermodynamic properties of surface silanol groups. First, we constructed model silica slabs via thermal annealing with reactive molecular dynamics and fully relaxed a number of candidate surfaces with accurate density functional theory calculations. We then dehydroxylated the surface by systematically removing neighboring pairs of silanol groups and calculated the energy change associated with each step. We used the resulting energetics in a thermodynamic model to compute the dehydroxylation temperature for each pair of silanols. Both the calculated total silanol concentration as a function of temperature and the relative concentrations of the different silanol types (isolated, geminal, vicinal) are in good agreement with experimental observations, validating our model.

(675f) Modeling Illite Within Shale: From a Density Functional Theory Perspective
Geatches, D.,
Wilcox, J., Stanford University
Originally presented on: 11/7/2013 14:10:00 - 14:30:00

MODELING ILLITE WITHIN SHALE: FROM A DENSITY FUNCTIONAL THEORY PERSPECTIVE

Dawn Geatches*, Jennifer Wilcox%

Department of Energy Resources Engineering, 367 Panama Street, Stanford University, Stanford, CA 94305-2220, USA: *geatches@stanford.edu

%wilcoxj@stanford.edu

The recent resurgence of interest in gas shale as a rich source of methane has raised many research questions at all length scales, from the seismic effects of hydrofracturing to interactions within the shale matrix at the atomistic level. Our research falls into the latter category, specifically investigating the interaction of clay minerals with H_2O, CH_4 and CO_2 within the wider research context of enhanced methane extraction and carbon dioxide storage and sequestration in exhausted shale beds. To identify possible chemical interactions between the clay mineral surfaces and H_2O, CH_4 and CO_2, we investigated the electronic structure of clay mineral models from first principles using density functional theory (DFT), plane waves and pseudopotentials. We focused on illite due to its prevalence within mature gas shale, and began by building the two main polytypes – 1M and $2M_1$ - based on crystallographic data, experimental formulae and illite-sample analyses. Illite (general formula $(Ca_{0.059}, K_{0.655})(Si_{3.597}, Al_{0.403})(Fe_{0.628}, Al_{0.969}, Mg_{0.428})$ per $O_{10}(OH)_2$) is classed as a dioctahedral interlayer-deficient mica characterized by non-swelling behavior and an interlayer dominated by potassium ions. Each layer consists of tetrahedral-octahedral-tetrahedral (TOT) sheets composed of (mainly) Si_2O_5 in the tetrahedral sheets and Al_2O_5 in the octahedral sheets. Substitutions of the Si^{4+} by Al^{3+} and octahedral Al^{3+} by Fe^{2+}/Fe^{3+} and/or Mg^{2+} produce a negatively charged TOT layer. The negative charge is balanced by positive interlayer cations, which, in the case of 1M illite, are mainly K^+, and less commonly Na^+ and Ca^{2+}. Within the interlayers of the $2M_1$ polytype it is also possible to find NH_4^+ and H_3O^+. The octahedral layer cations fill two of three possible octahedral sites, hence the description 'dioctahedral'. There are two types of occupation of these octahedral sites, the M1 and the M2 positions, which are identified by their vacancies and the positions of the hydroxyl groups around the vacant site.

Where pairs of hydroxyl groups lie on opposite apices of the vacant cation site this is `trans-vacant', and where the pairs lie on the same side of the vacant cation site this is `cis-vacant'. This short description of the possible variations in polytype, structure and composition of illite, explains why there were so few ab initio studies of illite available before we built our models.

In this presentation we show how, to account for these structural and compositional variations, we have explored cis- and trans-vacant structures of two main polytypes of illite found within shale and their various cation substitutions represented in the general formula. We also show how we described the strongly correlated d-electrons of iron, by testing a range of Hubbard (*U*) values within the DFT+*U* formalism, and when investigating clay mineral surface/H_2O/CO_2/CH_4 interactions we compared models with and without van der Waals forces, i.e. within the DFT+D formalism. Finally we interpret the nature of the surface/molecule interactions within the wider research area of enhanced methane extraction and carbon dioxide storage.

(709a) Computational Study of the Proton Affinity, 1 H NMR Chemical Shifts, and Thermophysical Properties of Imidazole Derivatives

Liu, H., University of Alabama
Bara, J. E., University of Alabama
Turner, C. H., University of Alabama
Originally presented on: 11/7/2013 15:15:00 - 15:36:00

There has been increasing interest in developing new materials for CO_2 capture. One of the approaches is to synthesize polymer membrane using imidazole derivatives as the monomer. These imidazole derivatives are able to react with aqueous CO_2 to form bicarbonate anions, thus enabling facilitated transport of CO_2 through the membrane. In order to better understand the chemical and thermophysical properties of imidazole derivatives, we have employed quantum mechanics and molecular dynamics methods to calculate the proton affinities, ^{1}H NMR chemical shifts, and absorption properties of a variety of imidazole derivatives. From our simulations, a quantitative characterization of the effect of various exocyclic substituents on the proton affinity of imidazole was achieved. In particular, we find that electron-donating groups are able to increase the proton affinity of imidazole. In addition, the calculated ^{1}H NMR chemical shift and thermophysical properties are in excellent with experiment. These results will be helpful for future experimental design of new materials to be used for CO_2 capture.

(709b) Separation Selectivity of Various Gases in the Ionic Liquid 1-Ethyl-3 -Methylimidazolium Tris(pentafluoroethyl)Trifluorophosphate

Althuluth, M., The Petroleum Institute
Peters, C. J., Petroleum Institute
Berrouk, A. S., Petroleum Institute
Kroon, M. C., Eindhoven University of Technology
Originally presented on: 11/7/2013 15:36:00 - 15:57:00

The separation of carbon dioxide (CO_2) from methane (CH_4) is an important process in many industrial areas such as natural gas processing and biogas purification [1]. Natural gas also contains significant amounts of ethane, some propane, butane, and other higher hydrocarbons [2]. In natural gas treating, loss of heavy hydrocarbons is a concern. It is desirable to recover these compounds due to practical problems.

There are many methods available for the removal of acid gases from gas streams. The most commonly used are chemical solvents, physical solvents, membranes and cryogenic fractionation [3]. Physical solvents tend to be favoured over chemical solvents when the concentration of acid gases is very high. However, if the concentration of heavy hydrocarbons is high, a physical solvent may not be the best option due to higher co-absorption of hydrocarbons. An acceptable solvent should have a high capacity for acid gas and a low capacity for hydrocarbons [4].

Therefore, ILs have been proposed to be used as a solvent for CO_2 capture, because of their advantageous properties over conventional ones [5]. We found that the ionic liquid (IL) 1-ethyl-3 methylimidazoliumtris(pentafluoroethyl) trifluorophosphate ([emim][FAP]) shows the highest carbon dioxide (CO_2) solubility of all ILs studied so far and shows higher selectivities for CO_2/CH_4 separations than any other IL, indicating the promising potential of using this IL for the separation of CO_2 from natural gas. Also, the solubilities of C_2H_6, C_3H_8 and C_4H_{10} in ([emim][FAP]) were measured and compared to the CO_2 solubility in the same IL.

The separation ratio between CO_2 and hydrocarbons decreases as the hydrocarbon chain becomes longer. The selectivity increase accordingly to the order: $S_{CO2/C4H10} < S_{CO2/C3H8} < S_{CO2/C2H6} < S_{CO2/CH4}$. A maximum selectivity is achieved at lower temperatures. Therefore, the CO_2 removal from a natural gas stream is recommended to be performed at low temperatures in order to achieve the best separation.

The obtained solubility data are modelled with the Peng-Robinson equation of state combined with quadratic mixing rule. The calculated data have been found to be in a good agreement with the experimental results.

Refrences:

[1] M. Arruebo, J. Coronas, M. MenÃ©ndez, J. Santamaría, Separation and Purification Technology 25 (2001) 275-286.

[1] R.W. Baker, K. Lokhandwala, Industrial & Engineering Chemistry Research 47 (2008) 2109-2121.

[3] G. Aldana, R. Arai, D. G. Elliot, An Evaluation of Sources of CO2 for EOR in Venezuela, Proceedings of the Sixty-Third Annual Gas Processors Association Convention, New Orleans, Louisiana, 1984.

[4] N. Korens, D. R. Simbeck, D. J. Wilhelm, Process Screening Analysis of Alternative Gas Treating and Sulfur Removal for Gasification, Prepared for U.S. Department of Energy by SFA Pacific, Inc., Revised Final Report, December 2002.

[5] E.D. Bates, R.D. Mayton, I. Ntai, J.H. Davis, Journal of the American Chemical Society 124 (2002) 926-927.

(709c) Modelling and Measuring Hydrate Promotion for CO2 Capture

Herslund, P. J.,
Thomsen, K., Technical University of Denmark
Abildskov, J., Technical University of Denmark
von Solms, N., Technical University of Denmark (DTU)
Galfre, A.,
Brantuas, P., Ecole Nationale SupÃ©rieure des Mines de St-Etienne
Kwaterski, M.,
Herri, J. M.,
Originally presented on: 11/7/2013 15:57:00 - 16:18:00

A novel separation concept is investigated here, whereby a precipitated, or slurry, system, enables CO_2 capture by hydrate formation – enhanced by thermodynamic promoters. We have demonstrated that a model for the hydrate phase, combined with an equation of state for associating systems – the Cubic Plus Association (CPA) equation of state enables an accurate determination of mixed hydrate dissociation pressures as well as hydrate composition.

We have subsequently presented experimental work measuring promoted gas hydrate formation with tetrahydrofuran (THF) and cyclopentane (CP), both known low-pressure/high temperature sII hydrate formers. However when applied as a mixed promoter system the hydrate formation pressures are lowered by up to 20% when compared to the pure CP promoter system, which is thermodynamically the more efficient of the two pure promoter systems.

Finally, in order to model this promoted system the complex phase behaviour between water and THF needs to be understood. This system is successfully modeled using CPA, including the complex LLE observed at high pressures. This complex system is best modeled by specifically accounting for the solvation of THF (i.e. that THF forms hydrogen bonds with water but is not self-associating), which is also the correct physical picture of what occurs.

(709d) Experimental Characterization Of Multiply-Substituted Alkylimidazoles As Solvents and Polymeric Membranes For CO2 Capture

Bara, J. E., University of Alabama
Shannon, M. S., University of Alabama
Moon, J., University of Alabama
Reclusado, K., Villanova University
Originally presented on: 11/7/2013 16:18:00 - 16:39:00

Previous work has characterized specific physical properties and CO_2 solubility of 1-*n*-alkylimidazole solvents for use in pre- and post-combustion CO_2 capture during gas treating. This research project explores the CO_2 solubility and physical properties of multiply-branched alkylimidazole solvents to determine the effects of functional group position(s) on various chemical and physical attributes. Using this information, additional correlations between ring substitution and physical properties can be derived to optimize the characteristics best suited for larger-scale pilot testing in industrial applications. Synthesis and purification of these compounds were followed by tests to determine density, viscosity, and vapor pressure as functions of temperature and ring substitution, and these properties were shown to be comparable to previously-tested 1-*n*-alkylimidazoles. The solvents in general were found to have low viscosities (<15 cP) and negligible vapor pressures (<0.5 mbar). CO_2 solubility experiments also showed that multiply-branched alkylimidazoles have similar if not improved absorption capacities relative to ionic liquids and 1-*n*-alkylimidazoles. In future work, polymeric membranes can be synthesized from these compounds to determine CO_2 permeability and selectivity as a function of ring substitution. The ultimate goal of this research is to find an imidazole-based solvent with low volatility and high CO_2 absorption capacity or an imidazole-based membrane structure with high CO_2 permeability and selectivity that can be easily manufactured for application in large-scale industry for the reduction of atmospheric CO_2 emissions.

(709e) Tuning the Physical and Chemical Properties of Novel CO2 Capture Solvents Based On Hybrid Nanomaterials

Petit, C., Imperial College London
Park, A. H. A., Columbia University
Originally presented on: 11/7/2013 16:39:00 - 17:00:00

Nanoparticle Organic Hybrid Materials (NOHMs) represent a novel class of liquid-like materials, whose potential applications include CO_2 capture. Their negligible vapor pressure and high thermal stability have indeed prompted the research community to investigate these nanomaterials as alternative carbon capture media to the common amine-based solvents. NOHMs are synthesized by grafting a nanosized inorganic core with a polymeric canopy. Prior studies have shed light on the mechanisms of CO_2 capture in NOHMs as well as their capture capacity and selectivity. In this study, the investigation of CO_2 capture using NOHMs is continued to investigate the effects of NOHMs' chemistry (*e.g.*, effect of amines functionalities) and structure (*e.g.*, effect of core size) on tuning the kinetic (*e.g.*, CO_2 diffusion), thermodynamic (*e.g.*, heat capacity), absorptive (*e.g.*, selectivity compared to H_2O and SO_2) and physical (*e.g.*, thermal stability) properties of NOHMs in terms of CO_2 capture. A range of analytical techniques are employed including various spectroscopical techniques (*i.e.*, ATR FT-IR, NMR, Raman), thermogravimetry and rheometry.

(709f) The Study of Using Novel Scrubbing Agent Mg(OH)2 Solution to Capture CO2 From Flue Gases

Cheng, L., University of Cincinnati
Li, T., University of cincinnati
Keener, T. C., University of Cincinnati
Lee, J. Y., University of Cincinnati
Originally presented on: 11/7/2013 17:00:00 - 17:21:00

Abstract

A novel CO_2 capture method by using environment-friendly chemical magnesium hydroxide solution has been tested and evaluated in response to minimizing the CO_2 emissions from coal-fired power plant in the post-combustion control area. The goal of this study is to conduct the technical feasibility studies in the area of physical &chemical characteristic of $Mg(OH)_2$, chemistry of the system, lean solution performance in absorption, capacity of regeneration in desorption. It is found that >90% removal efficiency can be achieved in the bubble column absorber; a C/Mg ratio of 9 is achievable at the condition of 52 ℃absorption and 75 ℃ desorption; a fresh feeding rate of 0.008 L/sec is optimized condition in the continuous test.

(709g) Electrochemical Strategies for Reduction of Greenhouse Gas Emissions

Stern, M. C., Massachusetts Institute of Technology
Herzog, H., Massachusetts Institute of Technology
Hatton, T. A., Massachusetts Institute of Technology
Eltayeb, A.,
Originally presented on: 11/7/2013 17:21:00 - 17:42:00

Anthropogenic carbon dioxide (CO_2) in the Earth's atmosphere has been cited as a primary cause of global climate change and threatens global public health and welfare. Carbon Capture and Sequestration (CCS) is an effective and important part of CO_2 emission abatement strategies, with the major CCS efforts to date focusing on the removal of directly from large-scale carbon emitters and storing it in secure geologic reservoirs. Thermal-swing operations using aqueous base scrubbing followed by stripping at elevated temperature have been the chemical sorption processes most investigated over the past two decades for CO_2 capture. Considerable quantities of steam and heat are required to release the CO_2 after capture at low temperatures, and substantial parasitic energy losses result from the need to use excess steam and heat in order to meet the kinetic requirements of the process.

Electrochemically-mediated amine regeneration is a new post-combustion capture technology with the potential to exploit the excellent removal efficiencies of thermal amine scrubbers while reducing parasitic energy losses and capital costs. The improvements result from the use of an electrochemical stripping cycle, in lieu of the traditional thermal swing, to facilitate CO_2 desorption and amine regeneration; metal cations generated at an anode react with the amines, displacing the CO_2, which is then flashed off, and the amines are regenerated by subsequent reduction of the metal cations in a cathode cell. The advantages of such a process include higher CO_2 desorption pressures, smaller absorbers, and lower energy demands. Several example chemistries using different polyamines and copper are presented. Experimental results indicate an open-circuit efficiency of 54% (15 kJ/mole CO_2) is achievable at the tested conditions and models predict that 69% efficiency is possible at higher temperatures and pressures. A bench scale system produced 1.6 mL/min CO_2 while operating at 0.4 volts and 42% Faradaic efficiency; this corresponds to a work of less than 100 kJ/mole.

(747a) Modeling of Protein Crystal Aggregation in Batch Crystallization

Nayhouse, M., UCLA
Kwon, S., UCLA
Christofides, P. D., University of California, Los Angeles
Orkoulas, G., University of California, Los Angeles
Originally presented on: 11/7/2013 15:15:00 - 15:35:00

Protein crystallization plays a crucial role in the $1 trillion pharmaceutical industry and has significantly contributed to both the scientific advancement and the economic development. For example more than 100 therapeutic proteins have been licensed and a number of additional therapeutic proteins are currently under investigation. In the last few years, researchers have attempted to model protein nucleation (e.g., [1]-[2]) and crystals growth (e.g., [3]-[4]) and their efforts make it possible to quantitatively describe and even control the shape and size distributions of the protein crystals. Although there has been a number of computational and theoretical studies of understanding the aggregation of fine particles, no significant work has been done in the modeling of protein crystal aggregation due to stirring in batch crystallization. Besides, no systematic approach has been established that visualizes the dependence of the aggregation mechanism of protein crystals on the temperature in the crystallizer and the solute concentration in the continuous phase where such information can be utilized to model and operate industrial-scale batch protein crystallization processes. In our previous works [5]-[6] batch crystallization of lysozyme accounting for crystal nucleation and growth was studied and a model predictive control strategy was proposed to regulate crystal shape and size distributions.

The present work focuses on the modeling of aggregation of protein crystals in addition to accounting for crystal nucleation and growth. First, we assume that the continuous phase is dilute enough to make only binary aggregation possible. In general crystallization experiments at higher stirrer speeds exhibit a reduced formation of crystal aggregates, which is favorable in the subsequent separation process by filtration where there is a physical limitation on the stirrer speeds that is achievable by a motor. Therefore an optimal hydromechanical stress is chosen, which enables the crystals to be maintained in suspension in industrial particulate processes [7]. Furthermore, the corresponding turbulent shear rate within the crystallizer can be characterized by the average velocity gradient of the flow field. An appropriate aggregation kernel is chosen which represents the rate at which a binary collision occurs and this rate strongly depends on the size of crystals as well as the crystallizer operating parameters [8]-[9]. Additionally, we assume that the aggregates of crystals are compact because they are subject to strong shear flow, and the corresponding collision efficiency to the impermeable aggregates is modeled [10]. Among the several physical behaviors that can induce particle aggregation, the aggregation for lysozyme crystals with a diameter approximately in the range 1-100 micrometer is mainly induced by shear forces according to its Kolmogorov microscale [9]. Lastly, an aggregate will be formed as two crystals completely merge along with their internal coordinates resulting in a decrease in the total number of crystals and an increase in the average crystal size. The subsequent shape changes in those crystals are accommodated by a set of internal coordinates into (110) and (101) directions. Extensive simulation studies are carried out to evaluate the influence of crystal aggregation on the shape and size distributions of the produced crystals at the end of a batch.

References:

1. Galkin O, Vekilov PG. Direct determination of the nucleation rates of protein crystals. J. Phys. Chem. B. 1999;103:10965-10971.
2. Pusey ML, Nadarajah A. A model for tetragonal lysozyme crystal nucleation and growth. Crystal Growth & Design. 2002;2:475-483.
3. Durbin SD, Feher G. Crystal growth studies of lysozyme as a model for protein crystallization. J. Cryst. Growth. 1986;76:583-592.
4. Forsythe EL, Nadarajah A, Pusey ML. Growth of (101) faces of tetragonal lysozyme crystals: measured growth-rate trends. Acta Cryst.D. 1999;55:1005-1011.
5. Nayhouse M, Kwon S, Christofides PD, Orkoulas G. Crystal shape modeling and control in protein crystal growth. Chem. Eng. Sci. 2013;87:216-223.
6. Kwon S, Nayhouse M, Christofides PD, Orkoulas G. Modeling and control of protein crystal shape and size in batch crystallization. AIChE J. in press.
7. Henzler HJ. Particle stress in bioreactors. Biochem Eng Biotechnol. 2000;67:35-82.
8. Saffman PG, Turner JS. On the collision of drops in turbulent clouds. J. Fluid Mechanics. 1954;1:448-462.
9. Kusters KA, Wijers JD, Thoenes D. Aggregation kinetics of small particles in agitated vessels. Chem. Eng. Sci. 1997;52:107-121.
10. Vanni M, Baldi G. Coagulation efficiency of colloidal particles in shear flow. Adv. Colloid Interface Sci. 2002;97:151-177.

(747b) Discovering Crystals Using Shape Matching and Machine Learning

Phillips, C. L., University of Michigan
Voth, G. A., Center for Biophysical Modeling and Simulation, University of Utah
Originally presented on: 11/7/2013 15:35:00 - 15:55:00

As the effective interactions between particles at the nano- to colloidal scale can be tailored and tuned in many ways, by using DNA and polymer functionalization, depletants, and solvents, there is a vast parameter space of design choices. Using simplified coarse-grained representations of the effective interactions between particles, computer experiments are an essential research tool for both exploring the vast space, and for investigating the first principles of self-assembly. The rate at which data can be amassed through computational simulation continues to accelerate, and thus the pace of discovery becomes limited not by the rate at which data can be generated, but can be analyzed.

We show how new crystalline structures can be identified automatically from analysis of large data set. By deploying a hierarchy of pattern analysis techniques using shape matching[1] and machine learning algorithms[2], local structures are extracted, classified, and then used to partition a data set into a phase diagram of similar crystals[3]. This method requires no a-priori knowledge of what might be present in the data set.

To demonstrate the method, we apply it to data generated from a parameter sweep of a constructed pair potential for a system of interacting particles (though our approach is generalizable to any model or data source). The data set contains a broad range of structures from simple to complex. We show how phase diagrams can be automatically generated, searches for new structures focused, and how the discovery of new materials can be accelerated.

[1] C.R. Iacovella, A.S. Keys and S.C. Glotzer, "Characterizing complex particle morphologies through shape matching: Descriptors, applications, and algorithms," J. Comp. Physics **230**(17), 6436-6463 (2011)

[2] A. Pandini, A. Fornili and J. Kleinjung, "Structural alphabets derived from attractors in conformational space," BMC Bioinformatics **11** (2010)

[3] C.L. Phillips, G.A.Voth, "Discovering crystals using shape matching and machine learning," Preprint, 2013

(747c) Current-Driven Evolution of Single-Layer Epitaxial Islands and Island Pairs On Crystalline Solid Substrates

Dasgupta, D., University of Massachusetts Amherst
Sfyris, G. I., University of Massachusetts Amherst
Maroudas, D., University of Massachusetts Amherst
Originally presented on: 11/7/2013 15:55:00 - 16:15:00

The driven assembly of confined quantum structures is of special importance to nanoelectronics and nanofabrication technologies. In this context, a particularly interesting problem is the current-driven dynamical response of single-layer adatom and vacancy clusters, i.e., islands and voids of single-layer thickness/depth, on substrate surfaces. In this presentation, we report theoretical and computational results on the current-driven morphological response of single-layer epitaxial islands and island pairs on crystalline elastic substrates based on a two-dimensional fully nonlinear continuum model, according to which mass transport due to curvature-driven diffusion, stress-driven diffusion, and electromigration is limited only to the island periphery (edge). The model also accounts for diffusional anisotropy for such "edge diffusion". Within this edge-diffusion driven mass transport regime, we have carried out a theoretical analysis of the morphological stability of such isolated islands and of the current-driven migration dynamics of morphologically stable islands. Employing well-tested front tracking methods, we have also developed direct dynamical simulators of the driven morphological evolution of such islands based on the model. We have validated the model by comparing the simulation predictions for stable island morphology and island-size dependence of the stable island migration speed with recent experimental measurements of current-driven evolution of homoepitaxial islands on Ag(111) substrates based on *in situ* scanning tunneling microscopy.

We have explored systematically the migration dynamics of morphologically stable steady islands over a broad range of misfit strains and kinetic parameters and found that the islands' driven translational speed v_m is inversely proportional to their size R_s up to a critical island size. For larger-than-critical sizes, the $v_m(1/R_s)$ relation becomes nonlinear and the island dynamics in the nonlinear regime is characterized by morphological transitions (such as faceting and fingering transitions), Hopf-bifurcation transitions to oscillatory asymptotic states, and morphological instabilities (such as necking instabilities). We derive a universal linear relationship that can describe this complex nonlinear behavior through rescaling v_m with the proper island morphological metric.

We have also analyzed the current-driven dynamics of single-layer epitaxial island pairs, where the two islands of the pair are driven toward their contact, and explored the resulting coalescence and breakup phenomena and the complex patterns they facilitate on the surface. Important geometric parameters in the analysis include the sizes of the two islands in the pair and the misalignment angle formed by the applied electric-field direction and the line connecting the centers of mass of the two islands in the initial configuration of the pair. We have discovered dynamical responses that lead to steady states after the initial island coalescence event. Depending on the initial island sizes, following coalescence, the resulting larger island may either form a single-island steady state or break up. If a single steady island is formed, its size determines whether its migration speed is in the linear or in the nonlinear regime. If island breakup happens following coalescence, the distribution of the resulting islands with respect to the applied electric-field direction is symmetric for aligned islands and asymmetric for misaligned islands. The findings of this work set the stage for systematic computational and experimental studies of external-field-enabled surface nanopatterning.

(747d) Protein-Sorting, Curvature-Induction and Curvature Sensing in Lipid Membranes: Quantification of Free Energies in a Mesoscale Cell Membrane Model

Tourdot, R., University of Minnesota
Natesan, R., University of Pennsylvania
Radhakrishnan, R., University of Pennsylvania
Originally presented on: 11/7/2013 16:15:00 - 16:35:00

Background: Curvature inducing cell membrane proteins are an essential component of many cell membrane remodeling processes including cell motility and internalization of proteins/cargo through endocytosis. Membrane proteins such as Epsin and Ampiphysins have been shown to tubulate liposomes in vitro, and migrate along mean curvature gradients in vitro [1, 2]. This curvature-mediated recruitment may play a role in vivo during Epsin recruitment to a nascent endocytic budding. Membrane proteins are thought to induce curvature through either the insertion of a amphipathic helix, or electrostatic scaffolding [3,4,5]. Some proteins induce curvature through a combination of these mechanisms [6]. It is not clear how these proteins sense mean curvature gradients with different affinities, and what major determinants impact their segregation. Some of the physical properties and experimental conditions that are relevant to the curvature-induction and sensing process are the magnitude and extent of the induced curvature by a single protein on the membrane, the membrane bending stiffness and background curvature, interfacial tension etc.

Methods: In this work, the lipid/cell membrane is modeled with a continuum Monte Carlo model evolved with the Helfrich Hamiltonian [7, 8]. In the model the membrane is described by a triangulated sheet that can access length scales on the order of 100 nm. Membrane proteins interact with the membrane through the intrinsic curvature field, and can move along the triangulated mesh according to Metropolis. The curvature field is approximated as a Gaussian, with Gaussian parameters derived from relating the curvature energy to isothermal titration experiments measuring Epsins association energy with the bilayer. Membrane proteins thought to induce an anisotropic curvature field are modeled by an anisotropic two-dimensional Gaussian function.

The Widom particle insertion method is a free energy calculation developed to compute a system's chemical potential [10]. Widom insertion probes chemical potential by randomly inserting a ghost particle, and determining the energy difference due to the extra particle. This method is only efficient for systems at low-moderate density. We employ the Widom insertion method to inhomogeneous systems where the excess chemical potential is computed along an intrinsically inhomogeneous axis of the system.

Results: Using our mesoscale membrane model, the chemical potential of curvature inducing proteins is quantified for two in vitro systems: (1) cylindrical lipid tethers (20-50 nm diameter) pulled from giant unilamellar vesicles (GUVs), and supported bilayers placed atop fabricated wavy substrates [2,9].

In order to compare our results to these in vitro experiments three model geometries were simulated, namely, a plane, a tube, and a sinusoid. The tube geometry approximates a membrane tether, while the sinusoidal surface is analogous to curvature sensing experiments on wavy substrates. All geometries are subject to periodic boundary conditions. The effect of random pinning to maintain the overall membrane topology is also investigated. A pinning study shows that as pinning density increases and membrane fluctuations are suppressed, the chemical potential of a curvature inducing protein increases nonlinearly.

In the planar geometry, the chemical potential required to bend the membrane to a certain curvature field is shown to depend on both the extent as well as the amplitude of the curvature field. A larger curvature field senses mean curvature gradients to a greater degree. The chemical potential also scales linearly with bending rigidity. In the tube geometry, there is a constant background mean curvature, as the radius of the tube increases, the mean curvature decreases, and the chemical potential increases. This finding is consistent with experiments performed on lipid tethers, which show a linear increase in fluorescence intensity with decreasing radius.

In the sinusoid geometry there is a gradient of mean curvature, this allows calculation of the density scaling of membrane proteins with mean curvature through inhomogeneous Widom insertion method. Experiments on wavy substrates show changes in density scaling for several different curvature inducing proteins. Simulations capture different scaling behavior according to a range of different Gaussian curvature field parameters. Anisotropic curvature fields are found to prefer parallel alignment with mean curvature gradients. This result is found by generalizing inhomogeneous Widom insertion along both the axis of the sine wave, and the angle of the curvature inducing protein. Simulations calculate lower chemical potential when the curvature field best matches the background mean curvature field.

In summary, the chemical potential of a curvature inducing protein in the dilute limit is shown to be dependent on the form and strength of the curvature field, the bending rigidity, and thermal fluctuations. For an anisotropic curvature inducing protein, the chemical potential shows preferential alignment/orientational favorability over spatial gradients in mean curvature. Our results are also shown to be consistent with in vitro experiments of protein segregation on curvature fields.

References:

[1] Ford, M., et al. Curvature of clathrin-coated pits driven by epsin. Nature. 2002.
[2] Capraro, B., Yoon, Y., Cho, W., Baumgart, T. Curvature Sensing by the Epsin N-TerminalHomology Domain Measured on Cylindrical Lipid Membrane Tethers. J. Am. Chem. Soc. 2010.
[3] Campelo, F., McMahon, H., and Kozlov, M.The Hydrophobic Insertion Mechanism of Membrane Curvature Generation by Proteins. Biophysical Journal. 2008.
[4] McMahon, H., and Gallop, J. Membrane curvature and mechanisms of dynamic cell membrane remodeling Nature. 2005.
[5] Zimmerberg, J., and McLaughlin, S. Membrane Curvature: How BAR Domains Bend Bilayers Current Biology. 2004.
[6] Fernandes, F., et al. Role of Helix 0 of the N-BAR Domain in Membrane Curvature Generation. Biophysical Journal. 2008.
[7] Helfrich, W. Elastic Properties of Lipid Bilayers - Theory and Possible Experiments. Z.Naturforsch. 1973.
[8] Ramakrishnan, N., Kumar, P. B. and Ipsen, J. H. Monte Carlo simulations of fluid vesicleswith in-plane orientational ordering. Phys. Rev. E. 2010.
[9] Hsieh, W. et al. Curvature Sorting of Peripheral Proteins on Solid-Supported Wavy Membranes. Langmuir. 2012.
[10] Widom, B. Some Topics in the Theory of Fluids. J. Chem. Phys. 1963.

(747e) Multiresolution Modeling of Polymer Solutions: Wavelet-Based Coarse-Graining and Reverse-Mapping

Agarwal, A., RWTH Aachen University
Adorf, C. S.,
Iacovella, C. R., Vanderbilt University
Ismail, A. E., RWTH Aachen University
Originally presented on: 11/7/2013 16:35:00 - 16:55:00

In contrast to multiscale methods, which can encompass multiple simulation techniques, *multiresolution* models uses the same general modeling technique with representations of a system at different length and time scales. We present work on a "complete solution" for modeling of semidilute polymer solutions, containing both coarse-graining and reverse-mapping steps.

The coarse-graining approach is based on wavelet-accelerated Monte Carlo (WAMC) method, which forms a hierarchy of resolutions to model polymers at length scales that cannot be reached via atomistic or even "standard" coarse-grained simulations. Using multiple stages of resolution, it is possible to simulate polymers of up to millions of repeat units in length. Although previously applied only to individual chain statistics, we show here how it can be extended to the study of polymer solutions. Bonded and non-bonded potentials between coarse-grained superatoms are computed. The non-bonded potential is computed using the same approach used previously in the study of single chains while the bonded potentials are computed using reverse Monte Carlo. A universal scaling function is obtained so that potentials do not need to be recomputed as the scale of the system is changed. To model polymer solutions, the intermolecular potential between the CG beads is assumed to be equal to the non-bonded potential, which is a reasonable approximation in case of semidilute systems. Using these potentials, various radial distribution functions are calculated from the coarse-grained representation at different resolutions, which compare favorably with the results from the atomistic simulations. We show that coarse-grained polymer solutions can reproduce results obtained from the simulations of the more detailed atomistic system to a reasonable degree of accuracy.

The reverse-mapping proceeds in much the same way: using probability distributions obtained from the coarse-graining procedure with respect to bond lengths, angles, and torsions, as well as the non-bonded potentials, we can reconstruct a more detailed version of the polymer that is consistent both with geometric constraints as well as energetic considerations. Using a "convergence factor" within a Monte Carlo-based energy optimization scheme, we can successfully reconstruct entire atomistic configurations from coarse-grained descriptions. Applications for the reconstruction of atomistic models from united-atom models and detailed models from coarse-grained models will be shown.

(747f) Sequestering of Toxins for the Reversal of Drug Toxicity: A Coarse Grained Molecular Dynamics Study

Ayee, M. A. A.,
Akpa, B. S., University of Illinois at Chicago
Originally presented on: 11/7/2013 16:55:00 - 17:15:00

Eighty seven people die every day in the United States as a result of an unintentional poisoning. Of poisoning mortalities, more than ~90% are due to drug intoxications. *In vivo* experiments and rare clinical reports suggest that intravenous delivery of a lipid emulsion facilitates reversal of drug toxicity and rapid restoration of compromised cardiac function. Patients unresponsive to other resuscitation measures have regained a spontaneous and sustained cardiac output in as little as 60 seconds following a lipid intervention. However, the mechanism of action of this therapy is poorly understood. Consequently, it is not possible to predict which toxins would be susceptible to lipid therapy.

The most commonly cited mechanism of action – known as the 'lipid sink' – suggests that lipophilic toxins associate with lipid emulsion droplets via hydrophobic partitioning. However, many of the compounds in question exist in a predominantly charged state at physiological pH. Moreover, acidosis may occur during cardiac arrest, further reducing blood pH levels and increasing the degree of molecular protonation. Hence, electrostatic interactions likely play an important role in driving the sequestration of toxins. In this work, we aim to clarify the relative roles of hydrophobicity and charge in mediating the association of toxin molecules with lipid droplets.

Coarse grained molecular dynamics simulations utilizing the MARTINI force field are employed to develop molecular topologies for the multiphase system of interest (an aqueous/oil system with phospholipid monolayer interface). The MARTINI scheme has been shown to semi-quantitatively reproduce the fundamental structural and thermodynamic properties of biomacromolecules such as proteins, carbohydrates, and phospholipids. However, similar topologies for small molecules are not as well explored. Here, neutral and protonated coarse grained topologies are developed for several small molecules with varying levels of lipophilicity and differing degrees of protonation at physiological pH. These topologies are systematically developed based on the three-dimensional molecular structures of the toxins of interest. A parameterization scheme is developed and validated by computationally reproducing the experimentally determined octanol/water partition coefficients of toxin compounds. These topologies are then incorporated into the multiphase 'droplet' system for the purposes of our study.

(747g) Pattern Formation and Coarsening in Molecularly-Tethered Lipid Membranes

Abel, S. M., Massachusetts Institute of Technology
Originally presented on: 11/7/2013 17:15:00 - 17:35:00

Many biological processes involve the binding and subsequent spatial reorganization of complementary molecules on apposed membranes. Recent experiments have utilized short, laterally mobile DNA strands of various lengths to tether fluctuating and supported lipid bilayers, resulting in interesting patterns in which DNA strands of similar lengths colocalize in spatial regions on the length scale of microns [1]. We use theoretical and computational methods to study the dynamics of a fluctuating membrane that is tethered to a supported membrane by laterally mobile molecules. The shape of the fluctuating membrane is governed by a time-dependent Ginzburg-Landau model, which is coupled to reaction-diffusion equations describing the time evolution of the concentrations of molecules. The free energy of the system accounts for the surface tension and bending rigidity of the membrane, as well as molecular binding [2]. When two or more molecular species with different natural lengths are present, the species segregate into distinct spatial regions, with the local membrane separation accommodating the natural length of the molecules. Different patterns of segregation are observed depending on system parameters including the number of types of molecules, their natural lengths, and their relative concentrations. Interestingly, we find by analytical means that there is a driving force for molecular phase separation even in the limit of vanishing surface tension and bending rigidity. The model provides insight into recent experimental results, and we suggest additional experimental tests aimed at gaining additional insight into biological phenomena.

[1] Chung M.; Koo B.J.; Boxer S.G. *Farad. Discuss.* **2013**, 161, 333-345.

[2] Qi S.Y.; Groves J.T.; Chakraborty A.K. *Proc. Natl. Acad. Sci. U.S.A.* **2001**, 98, 6548–6553.

(748a) Invited Talk

Wu, D. T., Colorado School Of Mines

Originally presented on: 11/7/2013 15:15:00 - 15:45:00

This is the invited talk for this session.

This session focuses on multiscale simulation to characterize the behavior of polymeric materials and produce insight into structure property relationships. The appropriate simulation protocols span or connect two or more scales of time, size or detail in the polymeric materials. *** Presenters taking part in the Meet-the-Faculty-Candidate session should identify themselves when submitting an abstract.

This session focuses on multiscale simulation to characterize the behavior of polymeric materials and produce insight into structure property relationships. The appropriate simulation protocols span or connect two or more scales of time, size or detail in the polymeric materials. *** Presenters taking part in the Meet-the-Faculty-Candidate session should identify themselves when submitting an abstract.

This session focuses on multiscale simulation to characterize the behavior of polymeric materials and produce insight into structure property relationships. The appropriate simulation protocols span or connect two or more scales of time, size or detail in the polymeric materials. *** Presenters taking part in the Meet-the-Faculty-Candidate session should identify themselves when submitting an abstract.

(748b) Coarse-Grained Molecular Dynamics Simulations of Thermal Annealing of P3HT:PCBM Bulk Heterojunctions for Organic Photovoltaic Applications

Carrillo, J. M., Oak Ridge National Laboratory
Kumar, R.,
Goswami, M.,
Sumpter, B. G., Oak Ridge National Laboratory
Brown, W. M., Sandia National Laboratories
Originally presented on: 11/7/2013 15:45:00 - 16:05:00

Molecular dynamics simulation is important in the development of organic photovoltaics because it provides insights on the evolution and the factors affecting the morphology of the bulk heterojunction during the thermal annealing process. Lee, Pao and Chu [Lee *et al.*, Energy Environ. Sci.,2011, **4**, 4124-4132] recently developed a multiscale approach that allows for the simulation of blends of poly(3-hexylthiophene) (P3HT) and [6,6]-phenyl-C_{61}-butyric acid methyl ester (PCBM). Due to limitations in the system size and simulation duration, however, they were unable to obtain results representative of the entire thermal annealing process. Here we present results obtained on the Titan supercomputer for simulations that are 27 times larger and 10 times longer. The coarse-grained molecular dynamics simulations were performed up to 400 ns and the simulation sizes are comparable to the thickness of the active layer of a bulk heterojunction device (~100 nm). We present results detailing morphological changes in domain size and interfacial area that were not captured by the short simulations by Lee *et al.* We characterized and validated the force field with investigation of the P3HT persistence length, PCBM and P3HT packing, and P3HT chain alignment. Additionally, we further studied the effect of the degree of polymerization, *N* of P3HT on the morphology and investigated the miscibility of P3HT and PCBM.

(748c) Phase Behavior of Rod-Coil Block Copolymers in the Melt and At Fluid Interfaces

Wei, T., Lamar University
Riggleman, R. A., University of Pennsylvania
Originally presented on: 11/7/2013 16:05:00 - 16:25:00

Conjugated rod-coil block copolymers have attracted increasing interest in applications that include photovoltaics and organic electronics due to their novel supramolecular structures. These structures are a result of the coupling of microphase separation and liquid-crystalline ordering. However, due to the complex nature of the interactions between the polymer chains, the theoretical tools used to describe rod-coil block copolymers are not nearly as mature as those used to study flexible block copolymers. In this work, we have employed a multiscale modeling approach that includes atomistic and coarse-grained molecular dynamics as well as highly efficient Monte Carlo simulations to study the phase behavior of rod-coil block copolymers in the bulk phase and the solvent environments. Our Monte Carlo simulations capture the main characteristics of rod-coil block copolymers in the melt while fully incorporating the fluctuations in three dimensions. Our atomistic and coarse-grained modeling are used to determine the self-assembly behavior both in solution and at fluid interfaces, where we find significant differences compared to flexible, coil-coil block copolymers.

(748d) Mathematical Modeling and Optimization Of The Bulk Continuous Process For The Production Of High Impact Polystyrene

Spies, C., Instituto de Desarrollo Tecnologico para la Industria Quimica, INTEC (Universidad Nacional del Litoral - CONICET)
Berkenwald, E., Instituto Tecnologico de Buenos Aires, ITBA
Estenoz, D. A., Instituto de Desarrollo TecnolÃ³gico para la Industria Quimica, INTEC (Universidad Nacional del Litoral - CONICET)
Originally presented on: 11/7/2013 16:25:00 - 16:45:00

In this work, the modeling and optimization of an industrial bulk process for the High Impact Polystyrene continuous production is studied. The mathematical model for a generic reactor is based on the model of Estenoz et al. [Aiche J., 44 (2), 427-441, 1998] and it is extended to consider the complex kinetics associated to the use of initiator mixtures and the presence of high temperature reactions. The devolatilization stage considers the polymerization including the kinetics of oligomer formation and crosslinking reactions together with the interfacial mass transfer phenomena. The model allows the prediction of the molecular structure along the process and the melt flow index and swelling index of the final product. Theoretical predictions adequately reproduce the observed evolution of variables. Engineering applications of the plant model includes the optimization of steady state operation using a routine that is based on a Random Search Algorithm (RSA). This is a robust method that provides a relatively good solution in short simulation times.

(748e) Modeling the Dissolution of Crystalline Cellulose in Ionic Liquids

Rabideau, B. D., RWTH Aachen University
Ojeda, C., RWTH Aachen University
Ismail, A. E., RWTH Aachen University
Originally presented on: 11/7/2013 16:45:00 - 17:05:00

Cellulose, despite being energy-rich and in seemingly boundless supply, remains a major untapped source of raw materials, both for energy production as well as specialty manufacturing. In large part, this is a result of its natural resistance to breakup and dissolution as a result of its strong hydrogen-bond network. Though a handful of solvents, including a number of ionic liquids, have been identified as capable of dissolving cellulose, the precise attributes that enables them to function remains unclear. A better understanding of the breakup mechanisms and of cellulose-solvent interactions should aid in the search for newer and better solvents and lead to the ability to estimate dissolution times for natural cellulose fibers.

We performed MD simulations that systematically examine the breakup of small cellulose bundles in 12 different imidazolium-based ILs. We show what happens to the hydrogen bonding network of cellulose during this breakup and how the cellulose-solvent interaction energies can affect dissolution rates. Further, we propose an underlying mechanism that leads to this breakup. Using these new insights, we formulate a relatively simple kinetic model capable of describing the cellulose dissolution. We explore the main findings of this model and show how the insights obtained contribute to a fuller understanding of the dissolution process.

(748f) Multiscale Simulation of the Atomistically Tailored Polymers / Oligomers Films and Surface Characterization

Chung, P. S., Carnegie Mellon University
Park, S., Carnegie Mellon University
Jhon, M. S., Carnegie Mellon University
Originally presented on: 11/7/2013 17:05:00 - 17:25:00

Molecularly thin film materials have emerged as a highly interesting and useful functionality for the numerous sustainable nanotechnology applications ranging from protective and smart coatings to electronics and nano-mechanical systems, sensors, renewable energy systems, display technology as well as serving biological, analytical, and medical purposes. Since peculiar phenomena were observed by the coupling of functional groups (*e.g.*, dewetting), the control parameters of the molecular conformations and morphologies of polymeric / oligomeric films by tailoring the molecular structure and functionality are a key to achieving the enhanced film properties for the design of application-driven nano-components. In order to facilitate the analysis of experimental data and enable predictive engineering of such systems, solid theoretical understanding is necessary. In this study, we investigated the variation of molecular conformations and surface morphology of functional oligomers with atomistically tailored structure and functionality. Molecular dynamics simulation using a coarse-grained, bead-spring method was performed to generate the morphology from submonolayer to a few layer oligomeric films [1]. To characterize the surface morphology, we introduced the wavelet transform entropy [2] as well as correlation length and the roughness exponent. Different morphological states were found for various oligomers backbones, resulting from strong tethering of functional groups on the solid surface, and as a function of surface coverage. In addition, the correlation between the surface morphologies, transport properties, and the wavelet entropy will be given.

[1] P.S. Chung, *et al.*, *J. Appl. Phys.*, vol. **103**, p. 07F526 (2008).

[2] M.H. Bharatil, *et al.*, *Chemometr. Intell. Lab.*, vol. 72, p. 57 (2004).

(748g) Systematic and Simulation-Free Coarse-Graining of Polymer Melts Using Soft Potentials

Yang, D., Colorado State University
Wang, Q., Colorado State University
Originally presented on: 11/7/2013 17:25:00 - 17:45:00

Full atomistic or united atom simulations of multi-chain systems are not feasible at present due to their formidable computational requirements. Molecular simulations with coarse-grained models where each segment represents the center-of-mass of a group of atoms/monomers have to be used instead. While atoms or monomers interact with hard excluded-volume interactions (e.g., the Lennard-Jones potential) and cannot overlap, the coarse-grained segments can certainly overlap and should therefore interact with soft potentials that allow complete particle overlapping. Coarse-grained models, however, reduce the chain conformational entropy, which plays an essential role in the thermodynamic behavior of polymeric systems. In this work we use integral-equation theories and a relative entropy framework for coarse-graining to investigate how the soft potential varies with N (number of segments on each chain) and how well the coarse-grained models can reproduce both structural and thermodynamic properties of the original system. This will provide us with a quantitative basis for choosing small N-values that can still capture the chain conformational entropy, a characteristics of polymers.

AUTHOR INDEX

AUTHOR INDEX

AUTHOR INDEX

AUTHOR INDEX

AUTHOR INDEX

AUTHOR INDEX